HIGH ENERGY DENSITY AND HIGH POWER RF

Previous Proceedings in this Series of RF Workshops

Year		Held in	Publisher	ISBN
1998	4th	Pajaro Dunes, California, USA	AIP Conf. Proceedings Vol. 474	1-56396-104-0
1996	3rd	Hayama, Japan	not published	
1994	2nd	Montauk, New York, USA	AIP Conf. Proceedings Vol. 337	1-56396-408-2
1992	1st	Dubna, Russia	not published	

Other Related Titles from AIP Conference Proceedings

611 Superstrong fields in Plasmas: Second International Conference on Superstrong Fields in Plasmas
Edited by Maurizio Lontano, Gerard Mourou, Orazio Svelto, and Toshiki Tajima, April 2002, 0-7354-0057-1

606 Non-Neutral Plasma Physics IV: Workshop on Non-Neutral Plasmas
Edited by François Anderegg, Lutz Schweikhard, and Fred Driscoll, February 2002, 0-7354-0050-4

595 Radio Frequency Power in Plasmas: 14th Topical Conference
Edited by Tak Kuen Mau and John deGrassie, November 2001, 0-7354-0038-5

592 High Quality Beams: Joint US-CERN-JAPAN-RUSSIA Accelerator School
Edited by S. I. Kurokawa, S. Y. Lee, J. Miles, E. A. Perevedentsev, November 2001, 0-7354-0034-2

581 Physics of, and Science with, the X-Ray Free-Electron Laser: 19th Advanced ICFA Beam Dynamics Workshop
Edited by C. Pellegini, S. Chattopadhyay, M. Cornacchia, and I. Lindau, August 2001, 0-7354-0022-9

576 Application of Accelerators in Research and Industry: Sixteenth International Conference
Edited by J. L. Duggan and I. L. Morgan, July 2001, 0-7354-0015-6

572 Electron Beam Ion Sources and Traps and Their Applications: 8th International Symposium
Edited by Krsto Prelec, June 2001, 0-7354-0011-3

546 Beam Instrumentation Workshop 2000: Ninth Workshop
Edited by Kenneth D. Jacobs and R. Coles Sibley III, December 2000, 1-56396-975-0

To learn more about these titles, or the AIP Conference Proceedings Series, please visit the webpage **http://proceedings.aip.org**

HIGH ENERGY DENSITY AND HIGH POWER RF

5th Workshop on High Energy Density and High Power RF

Snowbird, Utah   1–5 October 2001

EDITOR
Bruce E. Carlsten
Los Alamos National Laboratory, Los Alamos, NM

SPONSORING ORGANIZATIONS
U.S. Dept. of Energy, Office of Science
High Energy Physics Division

Melville, New York, 2002
AIP CONFERENCE PROCEEDINGS ■ VOLUME 625

Editor:

Bruce E. Carlsten
MS H851, NIS-10
Los Alamos National Laboratory
Los Alamos, NM 87545
USA

E-mail: bcarlsten@lanl.gov

Authorization to photocopy items for internal or personal use, beyond the free copying permitted under the 1978 U.S. Copyright Law (see statement below), is granted by the American Institute of Physics for users registered with the Copyright Clearance Center (CCC) Transactional Reporting Service, provided that the base fee of $19.00 per copy is paid directly to CCC, 222 Rosewood Drive, Danvers, MA 01923. For those organizations that have been granted a photocopy license by CCC, a separate system of payment has been arranged. The fee code for users of the Transactional Reporting Service is: 0-7354-0078-4/02/$19.00.

© 2002 American Institute of Physics

Individual readers of this volume and nonprofit libraries, acting for them, are permitted to make fair use of the material in it, such as copying an article for use in teaching or research. Permission is granted to quote from this volume in scientific work with the customary acknowledgment of the source. To reprint a figure, table, or other excerpt requires the consent of one of the original authors and notification to AIP. Republication or systematic or multiple reproduction of any material in this volume is permitted only under license from AIP. Address inquiries to Office of Rights and Permissions, Suite 1NO1, 2 Huntington Quadrangle, Melville, N.Y. 11747-4502; phone: 516-576-2268; fax: 516-576-2450; e-mail: rights@aip.org.

L.C. Catalog Card No. 2002108568
ISBN 0-7354-0078-4
ISSN 0094-243X
Printed in the United States of America

CONTENTS

Preface vii
RF2001 Program Committee Members ix

NLC Klystron R&D 1
    G. Caryotakis
High-Power, Annular-Beam Klystron Amplifiers 15
    J. Pasour, D. Smithe, L. Ludeking, and M. Friedman
Klystron Life Results in Particle Accelerator Applications 21
    H. Bohlen
Recent Progress in Multi-Beam Klystron in IECAS 29
    D. Yaogen
Inductive Output Tubes—Status and Future Direction 35
    H. Bohlen
Recent Progress in Understanding the Physics of Plasma-Filled,
High-Power Microwave Sources 45
    G. S. Nusinovich, Y. P. Bliokh, T. M. Abu-Elfadl, A. G. Shkvarunets,
    D. M. Goebel, Y. Carmel, T. M. Antonsen, Jr., and V. L. Granatstein
Overview of LIGA Microfabrication 55
    J. Hruby
"Windowtron" RF Breakdown Studies at SLAC 63
    L. Laurent, G. Caryotakis, F. Glendinning, N. C. Luhmann, Jr., C. Pearson,
    G. Scheitrum, and D. Sprehn
RF Breakdown in High Vacuum Multimegawatt X-Band Structures 77
    V. A. Dolgashev and S. G. Tantawi
Active and Passive RF Components for High-Power Systems 83
    S. G. Tantawi and C. D. Nantista
Power Supply, Energy Storage Line, and Grid Pulsers for High
Voltage Gridded Klystrons 101
    R. F. Koontz
Development of an X-band RF Gun at SLAC 107
    A. E. Vlieks, G. Caryotakis, W. R. Fowkes, E. N. Jongewaard,
    E. C. Landahl, R. Loewen, and N. C. Luhmann, Jr.
Design of High-Power, MM-Wave Traveling-Wave Tubes 117
    B. E. Carlsten, L. M. Earley, W. B. Haynes, and R. M. Wheat
Theory and Experiment of Ultra High Gain Gyrotron
Traveling-Wave Amplifier 129
    K. R. Chu, T. H. Chang, L. R. Barnett, and S. H. Chen
Interaction Circuits for High Average Power Gyro-TWTs Based on
Monolithic Lossy Ceramics 141
    J. P. Calame, M. Garven, B. G. Danly, B. Levush, and K. T. Nguyen
140 kW W-Band Heavily Loaded TE_{01} Gyro-TWT Amplifier 147
    D. B. McDermott, H. H. Song, Y. Hirata, A. T. Lin, T. H. Chang,
    H. L. Hsu, K. R. Chu, and N. C. Luhmann, Jr.
New Opportunities in Vacuum Electronics Using Photonic
Band Gap Structures 151
    J. R. Sirigiri, C. Chen, M. A. Shapiro, E. I. Smirnova, and R. J. Temkin

**Progress Toward a Gigawatt-Class Annular Beam Klystron with a
Thermionic Electron Gun** 159
    M. Fazio, B. Carlsten, J. Farnham, K. Habiger, W. Haynes, J. Myers,
    E. Nelson, J. Smith, B. Arfin, A. Haase, and G. Scheitrum
Relativistic Magnetron with Diffraction Antenna 169
    M. I. Fuks and E. Schamiloglu
Review of Computational Models for High Power Microwave Sources 177
    E. M. Nelson

Agenda 187
Author Index 191

Preface

The RF 2001 Workshop at Snowbird, Utah, was the 5th Workshop in the series of workshops that began in 1992. The first of the series was held in Dubna, Russia (1992) and continued at Montauk, New York (1994), Hayama, Japan (1996), and Pajaro Dunes, California (1998). At RF 2001 there were approximately 40 attendees with most from the US. Representatives also attended from CERN and Taiwan. Attendees represented academic institutions, national laboratories, and industry. Unfortunately RF 2001 followed the tragic events of September 11th by less than three weeks. The impact on attendance was a reduction by almost a factor of two from RF 98. On a positive note, the smaller number of attendees coupled with the relatively isolated geographic location of the Wasatch Range had the effect of producing a more informal and intimate workshop atmosphere.

The topical areas covered by the workshop included klystrons, multi-beam klystrons, klystrinos, TWTs, other linear beam tubes including the inductive output tube and the magnicon, and plasma filled high power sources. Fast wave devices included the gyrotron amplifier, gyroklystron, and gyro-TWT. We discussed supporting technologies related to high power development that included electron guns, RF breakdown phenomena, vacuum windows, loads, and components including pulse compression systems.

RF 2001 convened 97 years after the discovery of the vacuum diode detector by John Ambrose Fleming in 1904. For a relatively mature field there is a staggering amount of new work in progress pursuing new directions for producing higher power and higher energy at higher frequencies. Three significantly new themes emerged from RF 2001. Sheet beams, which were more of a curiosity at RF 98 have now become mainstream with the plan by SLAC to incorporate the sheet electron beam into the klystron design for the Next Linear Collider. The multi-beam klystron has come of age and has become commercially available. New designs are underway for multi-beam klystrons at higher frequencies up to X-band. Finally, entirely new fabrication approaches such as LIGA microfabrication are being pursued for millimeter wavelength high power devices such as klystrinos and traveling wave tubes that offer the possibility of tight control of dimensional tolerances and surface finishes, along with the promise of economy of large-scale production.

A number of approaches are being pursued for higher power that include sheet beams, annular beams, multiple beams, pulse compression, and gridded klystrons. Encouraging experiments are underway already on many of these approaches. Much thorough work has been done in an attempt to understand high voltage RF breakdown that has eliminated many of the potential causes, but all we know at this time is how limited our understanding is about the causes of breakdown and pulse shortening. This is a difficult problem with many open questions that will hopefully continue to be aggressively pursued because of the broad impact across many classes of high power devices.

Supporting technologies play a crucial role in high power source development. Improvements in lossy materials and better understanding of the use of these materials has been an important contribution. The design of passive components appears to be limited only by the designer's imagination. Advances with three-dimensional codes are making possible designs that could only be dreamt about just a couple of years ago. Excellent R&D work is also occurring on high power electron gun technology in order to produce the intense beams for driving powerful RF sources.

Perhaps most impressive is all the cross-linking among industry, DOE labs, DOD labs, and universities. There is a spirit of collaboration and cooperation that can only serve to accelerate progress in the area of high power, high energy density components, devices, and systems.

The program for RF 2001 included 35 papers with 12 of these being invited. The format for RF 2001 was similar to RF 98 with discussion leaders stimulating discussion on selected topics. The continued success of this approach leads us to recommend the same approach for the next RF Workshop that will occur in the summer of 2003, tentatively on the eastern coast of the US.

We would like to express our sincerest gratitude to our staffs at Los Alamos and Davis who worked together as a highly effective team. Lynette Lombardo of UC Davis was our Workshop Coordinator. Her vast experience with workshop organization proved invaluable. Lynette handled all the contractual arrangements and financial aspects of the workshop that resulted in our maintaining solvency. Lisa Marie Rodriguez of Los Alamos was our Workshop Secretary. Lisa Marie helped with the workshop organization and proceedings editing, and handled all the conference materials, attendee lists, and the myriad of details that are always part of such an undertaking. We also would like to thank Jaime Kephart of Los Alamos who played a key role in helping us find such an incredible location for the workshop.

We also thank Dr. David Sutter, Chief of the Advanced Technology R&D Branch of the US Dept. of Energy Office of Science, High Energy Physics Division for the financial support that allows the publication of these proceedings. Dave has always been a strong supporter of R&D in the field of high power and high energy density RF. We also owe a debt of thanks to Bruce Carlsten who took on the task of editor for these proceedings. Finally, we want to thank the members of the technical program committee who are identified on the following page for all their help in pulling together such an interesting and stimulating technical program.

Michael V. Fazio *Neville C. Luhmann*
Los Alamos National Laboratory University of California Davis

RF 2001 Co-Chairmen

RF 2001 Program Committee Members

Heinz Bohlen	Communications & Power Industries
Bruce Carlsten	Los Alamos National Laboratory
Bruce Danly	Naval Research Laboratory
Dan Goebel	Boeing
Kyle Hendricks	Air Force Research Laboratory
Wes Lawson	University of Maryland
David M. Parks	DERA/Ministry of Defence, UK
Glenn Scheitrum	Stanford Linear Accelerator Center

NLC KLYSTRON R&D

G. Caryotakis

SLAC, 2575 Sand Hill Rd, Menlo Park, CA 94025, USA

Abstract. This is a progress report on two ongoing projects in the SLAC Klystron Department: The continuing development of XP-3, ("Designed for Manufacture", or "DFM") klystron, and a new program to develop a sheet-beam version of this NLC source, also producing 75 MW, 3-microsecond pulses, with a repetition rate of 120 Hz.

Fig. 1 is a picture of XP-3, which is currently being baked-out. It is an 8-cavity klystron, with 7 stainless steel gain cavities in a stainless steel drift tube, and a 5-cell copper output cavity. The periodic permanent magnet (PPM) beam focusing structure is external to the vacuum and clamps around the klystron drift tube and cavities. A section of this assembly is shown in Fig. 2. The magnets are made of NdFeB, and produce a peak field of 2500 Gauss. A beam "stick", i.e. a vacuum tube with an electron gun directing a beam to a drift tube without cavities, with identical optics and focusing structure as the final klystron, was built and tested in July. It was operated at the same voltage and current, and at the same pulse length and repetition rate as the final klystron. The results are shown in Fig. 3. Beam transmission was 99.9 per cent, through a drift tube whose length was 75% of that in the XP-3.

The XP-3 is the latest of a family of NLC klystrons. The first tubes (XL-4s) were designed for 50 MW, 1.5 µs pulses, and were focused in solenoid magnets. A total of 10 of these klystrons were produced and are currently used in experiments at NLCTA and the Test Lab. Following the XL-4, the need to eliminate solenoid power led to the development of XP-1, a 50-MW klystron, employing a beam focused with periodic permanent magnets (PPM). It was operated with 1.5 µs pulses at 55% efficiency, meeting all specifications. Drawings of this klystron were provided for industry to bid, and eventually two copies were produced and delivered to SLAC, by Toshiba in Japan, and Marconi in England. Unfortunately, it was not possible to evaluate either of these klystrons to date, since both were received with a very poor vacuum. The power for the next developmental klystron was raised to 75 MW, also with periodic focusing, in order to meet new NLC specifications, designed to reduce the number of tubes in the machine. This klystron, the 75XP-1, was operated with poor beam transmission because of magnet procurement problems, but produced, at a 5-Hz repetition rate, 75 MW at 2.9 µs. Finally, the "Designed for Manufacture" (DFM), or XP-3 klystron, was built as the "baseline" NLC klystron. It is designed to produce 75 MW at 3.2 µs and a repetition rate of 120 Hz. These latest specifications raised the average power of the NLC rf source to 30 kW and made the XP-3 not only

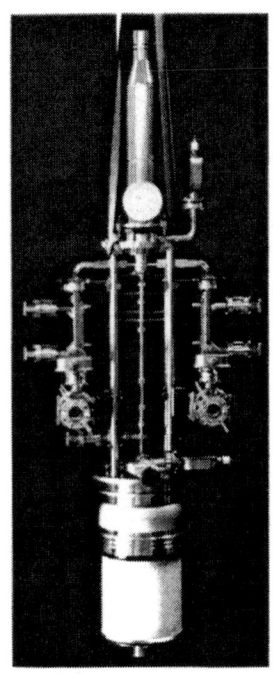

FIGURE 1. NLC DFM klystron.

FIGURE 2. PPM magnet assembly for DFM klystron.

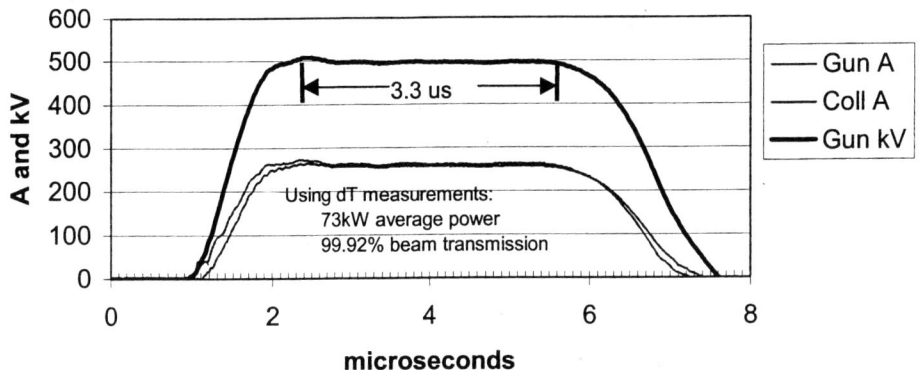

FIGURE 3. Beamstick results for DFM klystron.

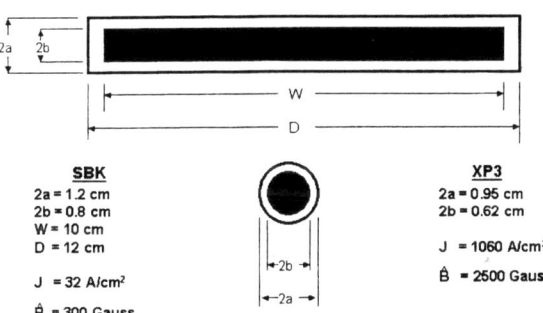

FIGURE 4. Comparison of round vs sheet beam parameters.

the most powerful peak power klystron ever developed at X-band, but also the highest average power X-band tube ever to be PPM-focused.

These XP-3 records have raised some concerns among SLAC klystron designers, because the power concentration in the drift tubes and cavities of the XP-3 may compromise yields in the test phase of tube production. For this reason, it was decided to investigate a sheet-beam klystron design addressing the same specifications, thus making it a "plug-compatible" NLC source.

Fig. 4 emphasizes the advantages of the present sheet-beam klystron SBK design (10-cm beam in a 12 cm-wide beam tunnel). Assuming that this SBK can be made to work, the selected beam parameters produce a 30-fold reduction in beam current density (and hence in power dissipation in drift tubes and cavities), and in a 8-fold reduction in the confining magnetic field. This design will also result in significantly lower manufacturing costs, as discussed later in this paper. Obviously, the assumption that such an SBK is feasible is one that must be substantiated before an experimental tube is constructed, since no one, anywhere, has ever built an SBK before, or even attempted to do so, to our knowledge.

There are several reasons for this. First, SBK design requires good 3-D simulation codes, which are available now, but which have not been in existence for very long. Secondly, prospective designers have been unnecessarily concerned with the stability of a sheet electron beam, because of "diocotron" instabilities that might cause the beam to break up. Thirdly, since the SBK drift tunnel can propagate all TE modes, the danger of feedback and oscillations is present.

Because the SBK design has gone through several stages of development, in which different beam voltages and sizes have considered, the examples cited below apply to various combinations of parameters, considered at various times. However, the examples used are valid illustrations of the general properties of SBK beams and circuitry. SBK parameters, at this stage of its design, are those shown in Fig. 4.

The simulations conducted to date, for both a sheet beam electron gun and the transport of the beam that issues from it, have produced conclusive proof that, for the parameter space used for this particular SBK, there is no difficulty in designing gun optics and beam transport. Fig. 5 is the gun design, the result of a 2D E-Gun simulation. An excellent beam is produced, with an area convergence of 10. Fig. 6 (3-D MAGIC) shows that the beam, with no confining magnetic field, spreads hardly at all in its wide dimension in 25 cm of distance (but eventually hits the wall in the narrow dimension); and that a peak magnetic field of less than 300 gauss (with no quadrupole component) provides perfect beam transport over the same distance (Fig. 7). We should add that for a diocotron instability to occur, a combination of both a much higher current density and a lower voltage would have been necessary; and, further, that such an instability could not occur in a periodic magnetic field of this period (3 cm).

We will now address SBK cavity design, together with the gain and stability. Fig. 8 is an early cold test model of cavity and drift tube, fabricated in two halves, cut at a symmetry plane at the midpoint and along the length of the beam. At the TM cavity mode of interest no currents cross this plane. The drift tunnel in this case is 9

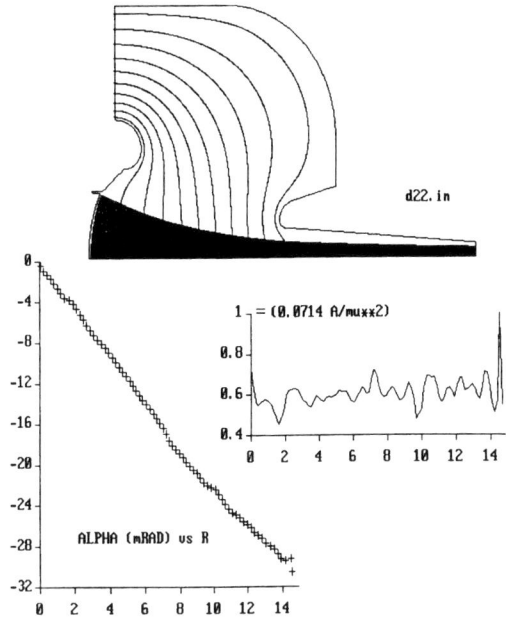

FIGURE 5. EGUN 2-D model of sheet beam gun.

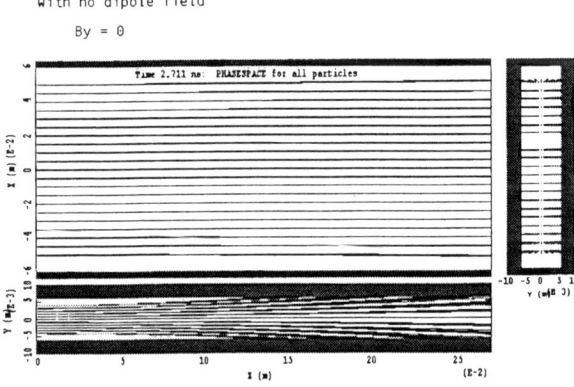

FIGURE 6. Magic model of beam transport with no magnetic field.

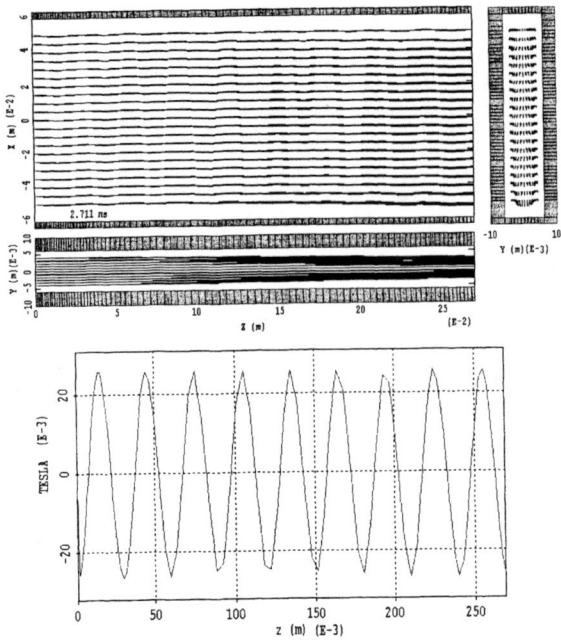

FIGURE 7. Magic 3D model of sheet beam with wiggler focusing.

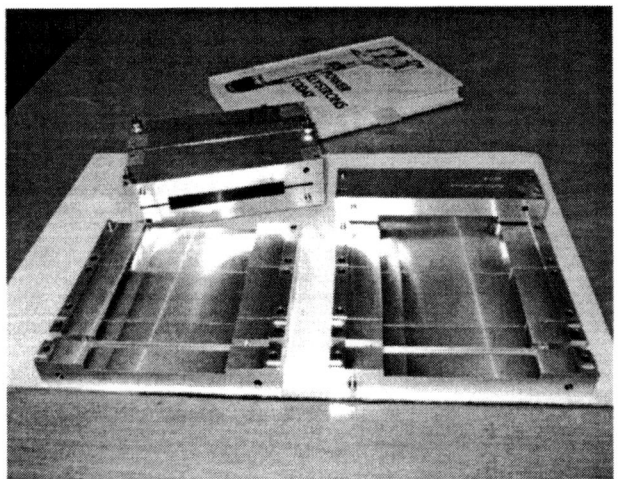

FIGURE 8. Cold test parts for sheet beam cavities and drift tubes.

cm wide and 1.2 cm high, and does not propagate TM modes below 12.65 GHz. TE modes do propagate, of course, and will be discussed later, when the stability of the device is explored. The cavities shown (half a cavity at the forefront and a full cavity at the end) consist of a single transverse cut below the drift tunnel surface, constituting half of a cutoff waveguide, which expands to a wider "a" dimension and terminates in two quarter-wavelength cavities (shown with symmetrical output couplings). An equivalent circuit for this type of cavity is shown in Fig. 9. The middle parallel circuits are all resonant at the same frequency (the cutoff frequency of the TE_{01} mode in the waveguide). The end quarter-wave cavities have different element values. Interaction with the rf- modulated beam is approximated by distributed current sources.

Because of its proportions and the size of the drift tunnel opening, the single cavity shown in Fig. 10 (the earlier case, with the 9-cm drift tunnel) produces a beam-modulating E-field, which leaks into the drift tunnel for a considerable distance, on either side of the cavity. As a result, the coupling coefficient (the convolution of the electric field from the gap with the cosine of the transit angle) is very low (the case in Fig. 10 applies to a narrower, 8-cm beam). This can be corrected if an extended 3-cell cavity is employed, with the cells coupled capacitively through the drift tunnel and synchronized with beam in a π-mode. It can be seen from this figure that the coupling coefficient increases from 0.09 to 0.36, going from one to three cells. The 3-cell R/Q is 54 ohms. The small-signal gain of the SBK is proportional to M^2 R/Q, with improves rapidly with the addition of more cells. Hence the use of multi-cell cavities is mandatory.

Since these cavities are overmoded in two dimensions, it is important to identify all TM modes and examine them for "monotron" oscillations, i.e. oscillations resulting from negative beam loading within the extended cavity. As TM modes will not propagate past the cavities, they cannot produce convective instabilities. Fig. 11 displays the TM modes near the operating frequency present in the 3-cell cavity of Fig.10. The beam loading each mode produces has been calculated using the method shown in Fig 12, where the difference in the squares of the coupling coefficients for the beam fast and slow waves is calculated from the cavity fields (at the beam midpoint). A negative difference indicates negative beam loading, and the possibility of monotron oscillation if the cavity is not sufficiently loaded [*].

TE modes are a different matter, since, if present, they will propagate in the drift tunnel, and could set up convective instabilities. Simulations will be conducted to establish whether reasonable fabrication tolerances can maintain sufficient cavity symmetry so that TE modes, which can trigger such instabilities, will not be excited.

---

[*] Fig. 12 plots the ratio of the beam-loading conductance Gb to the beam conductance Go, against beam voltage in kilovolts. The maximum negative value of Gb/Go is 0.035, (Go = 5.2×10^{-4} mhos). The beam-loaded Q is defined as Qb = 1/R/Q*Gb, and since R/Q for this cavity was calculated to be 25 ohms, it follows that, at approximately 250 kV, Qb = -2200 This indicates that the cavity will oscillate unless it is loaded, as would be the case if it were the output cavity. However, it could still be used as a "gain" cavity, if it were artificially loaded for a ohmic Q of about 2000. Shorter (3-cell) cavities display much lower values of negative beam loading.

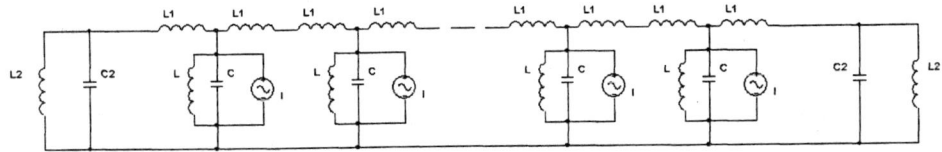

FIGURE 9. Equivalent circuit for single cell sheet beam cavity.

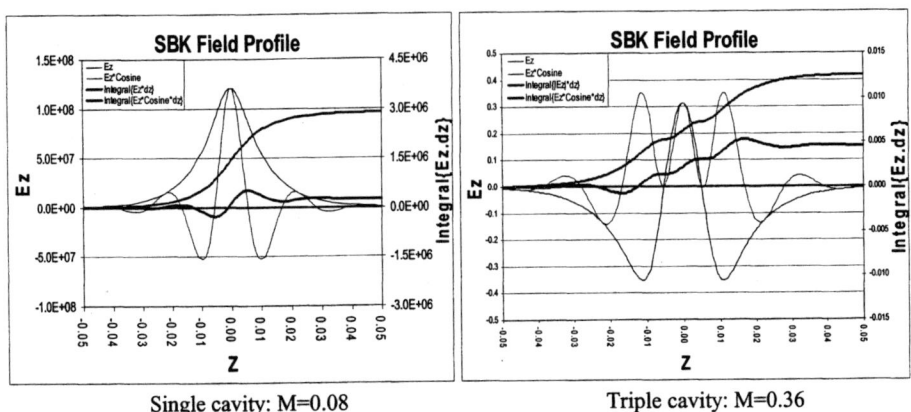

Single cavity: M=0.08   Triple cavity: M=0.36

FIGURE 10. Field profile and coupling coefficients for single and triple gap sheet beam cavities.

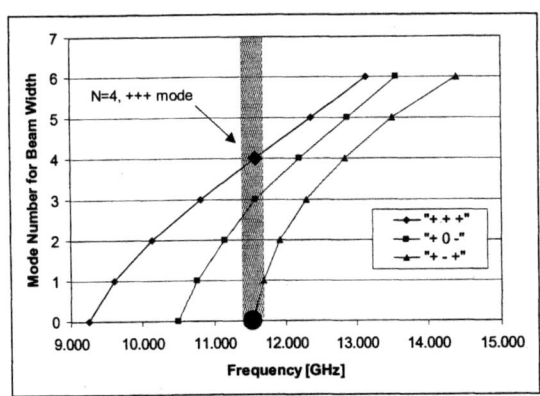

FIGURE 11. Mode plot for a 3-cell extended cavity ("+++"=2π, "+0-"=$\pi/2$, "+-+"=π)

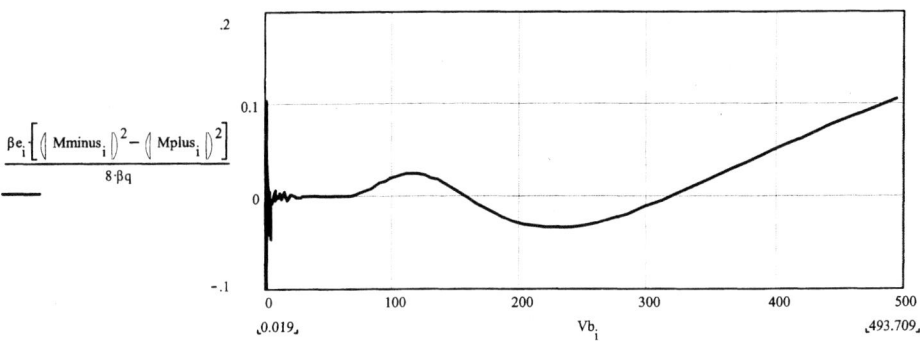

FIGURE 12. Plot of electronic conductance vs voltage.

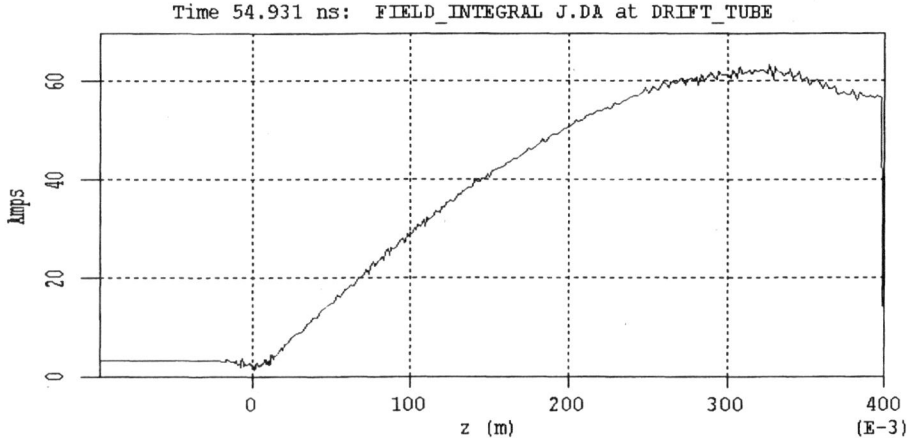

FIGURE 13. Current modulation from fields applied in a sheet beam cavity.

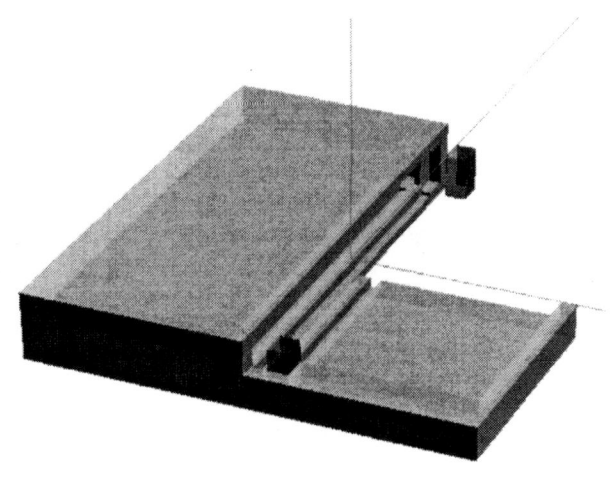

FIGURE 14. Solid model showing sheet beam cavity and drift tube.

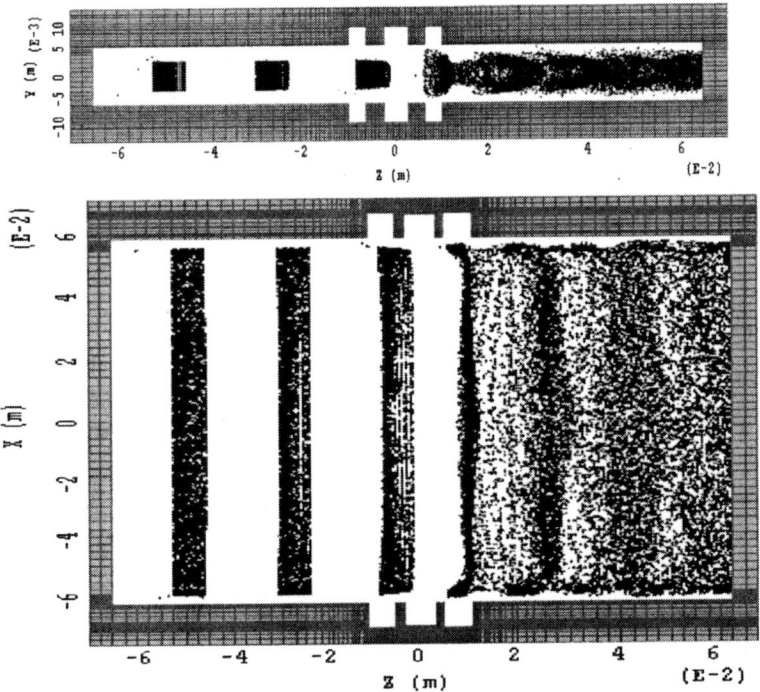

FIGURE 15. Side and top views of 3-D MAGIC simulation of beam propagation.

It has been determined by simulation that, at the current densities employed, beam non-uniformities will not set up TE mode monotron oscillations in extended cavities.

We are also building cold test models to determine whether lossy ceramics, embedded in the walls of the drift tunnels between cavities, can sufficiently dampen propagating TE modes to avoid potential convective oscillations; the same ceramics can also provide light loading at the leakage fields of the TM modes within the cavities, discouraging the onset of monotron oscillations, and providing more freedom in the design of extended cavities.

Finally, the small-signal and large-signal interactions between sheet beam and cavities have been investigated through MAGIC simulations. In Fig. 13, a small (2 A) rf current modulation was added to a 450 kV, 320 A beam, injected into the 3-cell cavity of Fig. 10. Measurements are made of the voltage developed across the cavity, as well as of the added rf current in the beam. The results agree with small-signal klystron theory [*].

A separate MAGIC simulation, using a 285A, 420 kV beam, with a 10 cm width, in a 12-cm tunnel, and a cavity with an R/Q of 18 ohms at the center, dropping off to 11 at the beam edges, produced the following large-signal results:

The DC beam was artificially modulated with an rf current equivalent to 1.8 times its DC current. This is the magnitude of the modulation expected in the actual SBK. The beam was injected into the cavity, which was equipped with artificial loads

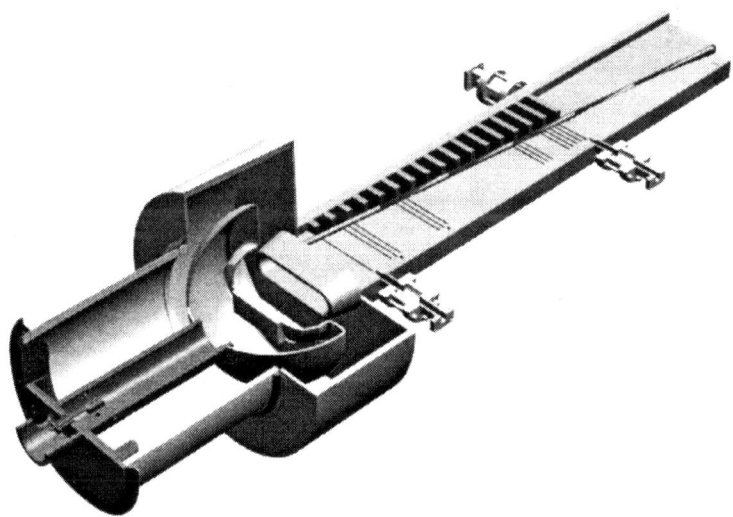

FIGURE 16. Solid model of sheet beam klystron.

---

[*] The additional current results as the velocity modulation imparted to the beam by the cavity voltage is converted to density modulation. Since the R/Q of the cavity is known, the voltage measurement provides information on the beam-loaded Q_b of the cavity (Q_b=860), while the current measurement at its maximum downstream of the cavity yields a value for the plasma frequency ω_q (reduced by image charges, ω_q is 1.2 GHz). The Q_b and ω_q, determined through this MAGIC simulation, are consistent with the calculation of these parameters through analytical methods.

at its 3rd cell providing a loaded Q of approximately 300 (Fig. 14). This simulation, without any adjustments, produced an output of 63 Megawatts into the loads, or an efficiency of 53%, despite the fact that the fields (or R/Q) at the cavity gaps were not uniform across the beam web. The current modulation in the beam before and after interaction is shown in Fig. 15.

The results of all the above simulations have been very encouraging. We are therefore proceeding with the fabrication of the "plug-compatible" SBK, shown in Fig 16. The advantages of this device, apart from the considerably enhanced average power capacity which motivated its design, are that a) compared to the DFM klystron,

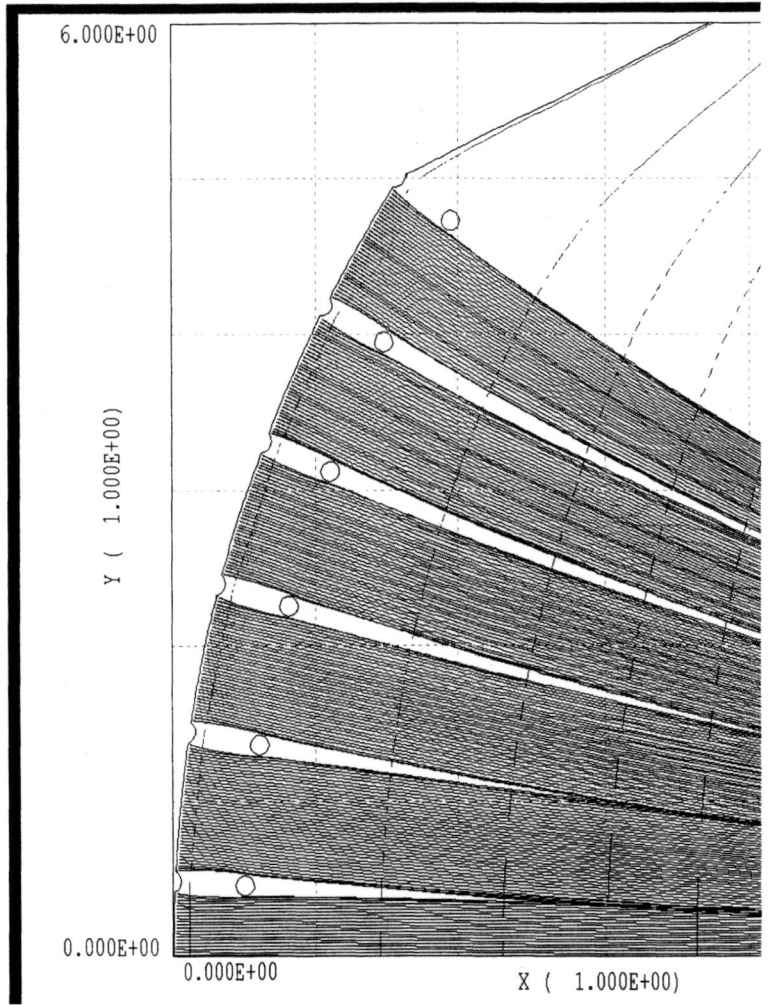

FIGURE 17. Grid close-up.

the parts count is greatly reduced, since cavities, collector and cooling passages can be fabricated on two copper plates on an CNC milling machine, b) the low magnetic field required makes the use of inexpensive ferrite magnets possible, and c) grid modulation is now practical, because of the sheet-beam gun topology, and the low beam current density. A gridded gun is currently being designed under a DOE SBIR with Calabasas Creek Research. Considerable progress has already been made in the design of this gun, as seen in Figs. 17 and 18. The total grid swing required (from cutoff to full current) is 8.5 kV in an SBK designed to produce 60 MW at a beam voltage of 415 kV.

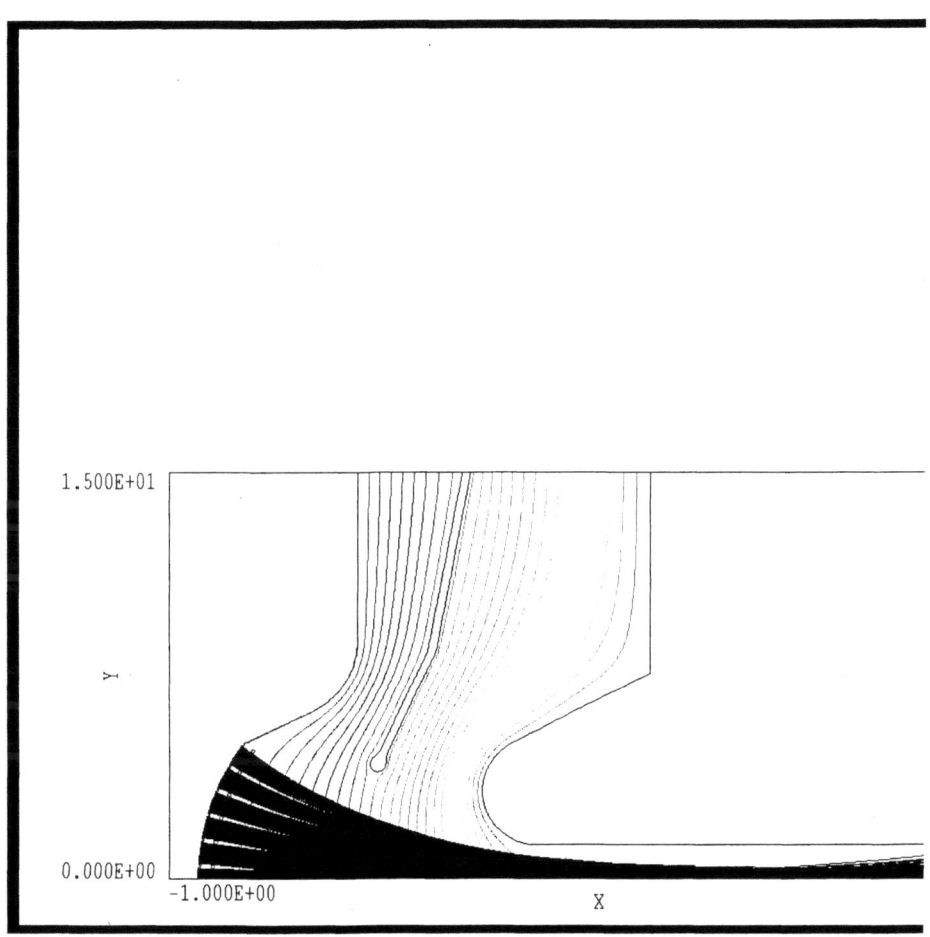

FIGURE 18. Gridded gun (with modulating anode)

High-Power, Annular-Beam Klystron Amplifiers

John Pasour[†], David Smithe[†], Larry Ludeking[†], and Moshe Friedman[*]

[†]*Mission Research Corporation*
8560 Cinderbed Rd., Newington, VA 22122
[*]*Naval Research Laboratory*
Washington, DC 20375

Abstract. Annular beam klystron amplifiers are being developed at L-band and at X-band. These devices are designed to operate at power levels of hundreds of MW to ~1 GW, with pulse durations up to 800 nsec. The L-band amplifier uses an 11-cm-diameter, 3-mm-thick annular beam (450 kV, 4.5 kA) inside an open beam tube with large-gap cavities. The X-band amplifier employs a 12-cm-diameter annular beam that propagates between inner and outer grounded cylinders and cavity structures. At higher frequencies or power levels, this so-called triaxial configuration provides a significant advantage over the open-cylinder configuration. In effect, it is a sheet-beam klystron bent into a full circle, thereby avoiding the edge effects. Alternatively, it can be thought of as the continuum limit of the multi-beam klystron.

INTRODUCTION

It is well known that the scaling of conventional klystrons to higher power, particularly at shorter wavelengths, is severely limited by electron beam propagation (space charge) issues. Thus, there has been increasing interest in configurations that allow higher beam current transport at modest beam voltages (500 kV or less), while still providing strong beam-cavity interactions and efficient operation. Sheet beam and multiple beam klystrons are examples of such configurations. An alternative approach that we have pursued for the past few years is based on a large-diameter, thin annular electron beam. This work has built on the pioneering research of Friedman et al. at the Naval Research Laboratory.[1,2]

The annular beam configuration has some significant advantages. At any given voltage, much higher beam current can be transported in an annular beam near the wall of an open tube than in a pencil beam. Adding a grounded drift tube and cavity structure inside as well as outside the beam further increases the current limit and reduces the magnetic field required for transport. Furthermore, unwanted TM modes in the drift regions between cavities can be cut off by making the spacing between the inner and outer tubes less than half a wavelength, regardless of the overall beam diameter. This feature greatly simplifies the scaling of high-power klystrons to higher frequencies.

Two separate annular beam klystron amplifiers will be described here. The first operates at 1.3 GHz and was motivated by a desire to achieve GW-level output for pulse durations of ~1 μsec. The second operates at 9.3 GHz and was motivated by RF accelerator needs. Both amplifiers use a CsI-coated graphite fiber cathode, which produces a uniform, constant-current, annular beam having a cross section of about 3 mm. The annular beam is propagated close to the drift tube wall in a uniform axial magnetic field of 2–3 kG. Beam voltage in both amplifiers is 400-450 kV, with current of 4-5 kA. The vacuum level is typically about 1×10^{-5} torr.

L-BAND AMPLIFIER

The L-band amplifier, shown schematically in Fig. 1, consists of three cavity structures connected by short sections of beam tube. The first two cavities are completely external to the beam tube. The third, or extraction, cavity is a triaxial design, with cavities and stacked annular disks placed both inside and outside the beam.

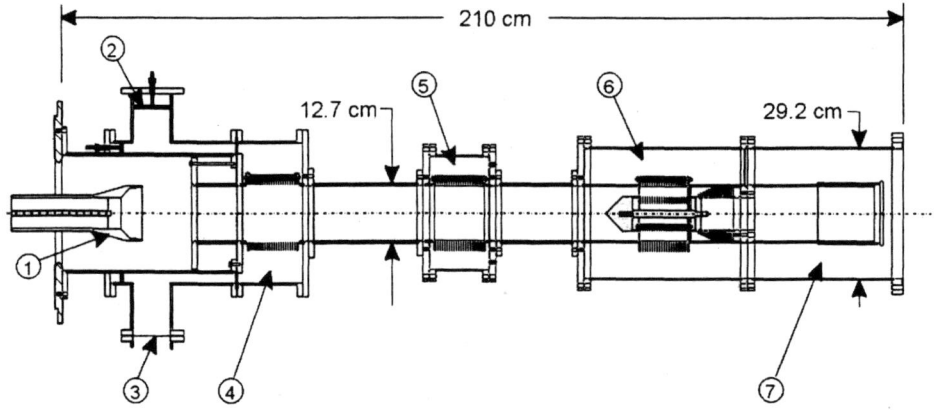

FIGURE 1. L-Band amplifier. 1. Cathode. 2. Tuner. 3. RF input. 4. Input cavity. 5. Idler cavity. 6. Output cavity. 7. TEM-TM_{01} mode converter.

All the cavity structures are wide-gap designs, with a stack of thin annular disks placed across the gap.[3] The disks are joined together by thin, axial rods. The rods form an inductive path and have little effect on the RF fields, but they appear as short circuits on the time scale of the beam. Thus, the effective conducting boundary for the beam is unbroken at the gap, eliminating the large increase in potential energy that would otherwise occur at such a wide gap and allowing a lower magnetic field to be used for transport. Furthermore, the inner radii of the washers taper down in the output gap, which restores even more of the potential energy to kinetic energy. Also, the tapering together with lower magnetic field allows any electrons that are reflected in the output gap to be collected at low energy on the washers. This removal of reflected electrons was proposed as an explanation for the observed increase in pulse length as the magnetic field was decreased in the initial experiments with wide-gap cavities.[2] Finally, the wide gaps allow operation at larger gap voltages without breakdown, which is an important advantage for longer pulse or higher power operation.

As shown in Fig. 2, the field across the gap is bidirectional. Thus, in operation the cavities perform somewhat like a π-mode multi-gap cavity. The gap length is chosen so that the electrons are properly phased with the field.

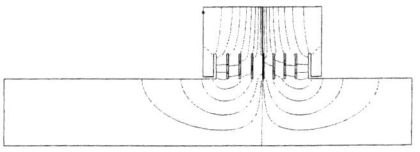

FIGURE 2. Superfish field plot of idler cavity.

MAGIC-2D and MAGIC-3D [4] simulations were performed to guide the design of the cavity structures. In particular, comparisons were made of the open-tube configuration and a triaxial configuration. At the 1-GW power level and at the L-band frequency of interest, we saw little benefit in the triaxial configuration except in the vicinity of the output gap. Therefore, we designed the system with a grounded structure inside the beam radius only in the region of the output cavity. This output cavity configuration bears some similarity to the RKO developed at AFRL by Hendricks, et al., [5] which also has a grounded center conductor in the output gap region. To take advantage of existing hardware and a load being separately developed by MRC, the RF output was extracted coaxially and injected into a circular waveguide in the TM_{01} mode.

Representative waveforms of the L-band amplifier are shown in Figs. 3 and 4. Figure 2 shows the injected beam current and cathode voltage as well as the modulated current downstream of the idler cavity. The gun voltage pulse is generated by a PFN/pulse transformer, with a high voltage PFN and a series spark gap on the output to provide pulse shaping. As such, the pulse has a sharp rise time, an 800 nsec flat top, and a rather long tail. Input power is provided by a 500-kW magnetron, which produces a maximum beam current modulation of about 8% at the location of the second cavity. At the extraction cavity, the bunching amplitude has increased to >75% of the injected current, as shown in Fig. 3. The modulated signal is produced by a heterodyned B-dot probe in the tube wall.

The output power is monitored using a 60 dB loop directional coupler borrowed from AFRL (Mike Haworth and Kyle Hendricks). This coupler is mounted between

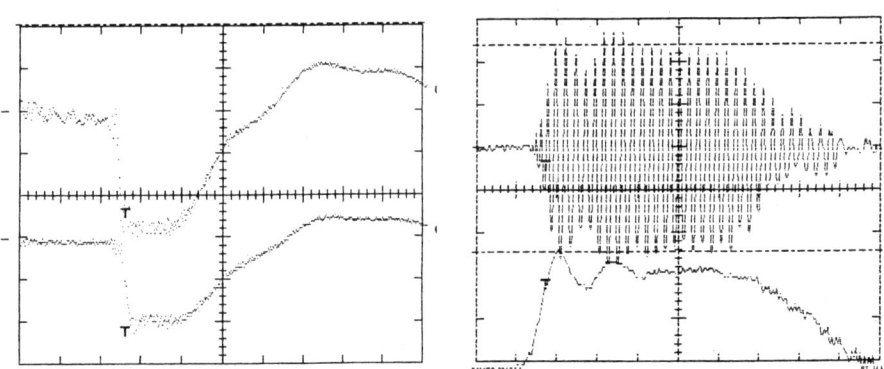

FIGURE 3. Left: Cathode voltage (top, 140 kV/div) and injected current (2 kA.div); 500 ns/div. Right: Modulated current after idler cavity (top, 1.5 kA/div) and cathode current (2 kA/div); 200 ns/div.

the TEM-to-TM$_{01}$ mode converter and the vacuum load. The coupler output is fed either to a calibrated crystal detector or a calibrated heterodyne detector through an attenuator stack. The crystal detector pulse shown in Fig. 4 indicates that the output pulse is relatively constant throughout the beam current pulse.

It is interesting to note that the output pulse does not change appreciably as a function of pressure from about 3×10^{-5} torr to 5×10^{-6} torr. At higher pressure, the pulse duration decreases. In fact, the beam modulation observed after either the input cavity or the idler cavity is markedly decreased at pressures above 5×10^{-5} torr.

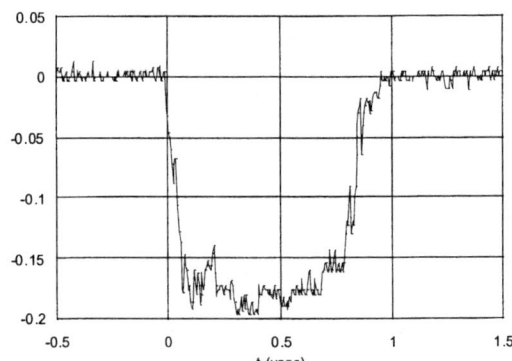

FIGURE 4. Crystal detector output signal of the L-band klystron. Peak power is ~500 MW.

X-BAND AMPLIFIER

An open-tube, coaxial configuration like the L-band klystron does not scale very well to higher frequencies. To cut off electromagnetic modes in the drift regions between cavities, the tube diameter would have to be reduced with the wavelength. In fact, Serlin and Friedman found [6] in limited tests that the output power of the coaxial klystron scaled roughly as λ^{-3}.

The triaxial configuration, with return current flowing both inside and outside the beam, overcomes the frequency scaling limitations of the coaxial version. The basic triaxial layout is shown in Fig. 5. The drive power is injected into the input cavity via a coaxial feed on the system axis. The configuration shown uses a single-gap input cavity and multiple-gap idler and extraction cavities. The central structure is supported and electrically connected to ground by rods that pass through large holes in the cathode stalk. This allows return current to flow at a radius less than the beam's while avoiding discontinuities in the beam tube (other than the cavity gaps). This approach provides much more stable beam transport than if the central structure is supported only from the output end or by spokes that pass

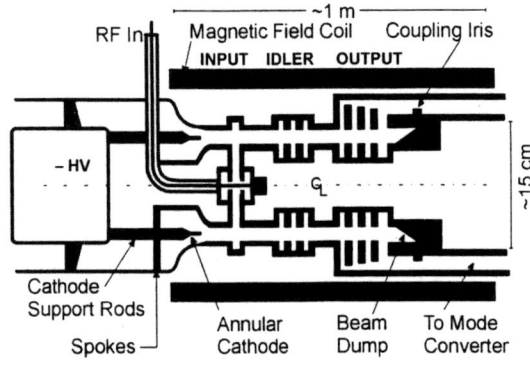

FIGURE 5. Schematic diagram of triaxial klystron.

through the beam. The cavities are designed to suppress TEM mode excitation in the drift region by introducing small asymmetries in the inner and outer cavities of each cavity pair. Small-diameter, axial support pins are placed at electric field nulls of the desired cavity mode to provide a return current path and suppress unwanted cavity modes.

SUPERFISH calculations of the input cavity and idler are shown in Fig. 6. The idler cavity and the output cavity use extended-interaction, inductively loaded gaps to allow higher-power operation. Typically, these are "π-mode" structures, with gaps spaced so that the bunched beam remains in phase with the decelerating field. The height of radial cavities must be an integral number of half wavelengths at the desired frequency, but the number of nodes in the radial direction can be chosen to adjust the ratio of energy stored to gap voltage. Great care has to be taken to ensure that competing modes are spaced well away from the desired mode and to minimize RF field leakage into the beam tunnel. Typically, proper design requires a careful balance of the inner and outer cavity wall radii.

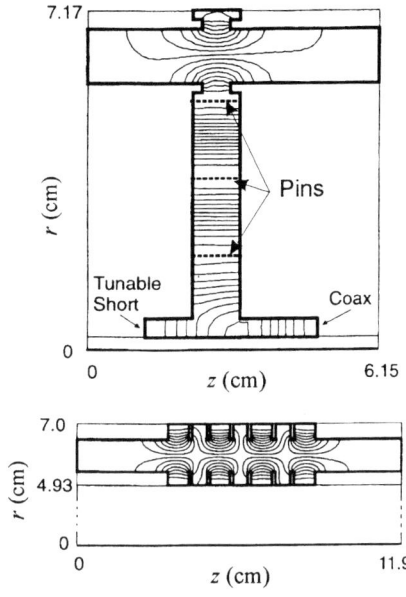

FIGURE 6. Top: Input cavity. Bottom: Idler cavity.

Stable beam modulation with an amplitude of ~10% of the injected current has been produced with the input cavity driven by a 9.3-GHz magnetron at a power input of ~180 kW, in very good agreement with numerical simulations. The modulation amplitude was very flat and persisted for the duration of the injected beam pulse, with excellent phase stability. Strong modulations could be produced over a bandwidth of 50 MHz simply by varying the drive frequency. With the idler structure placed about 6 cm from the input cavity, very strong modulation was observed within a few cm of the idler structure. With a 16-kA injected current, a maximum modulation amplitude of almost 10 kA (>19 kA peak-to-peak) was produced. Simulations show that bunching amplitude varies linearly with initial beam modulation amplitude, at least up to 20% initial modulation, with little variation in location of peak bunching. With proper frequency tuning, essentially total beam modulation can be produced by a 6-gap idler structure given a 10% initial beam modulation.

Computer simulations have shown 50% energy extraction in a multi-gap output structure. A four-gap structure of the type shown in Fig. 7 achieves this performance with a beam that has 75% modulation amplitude at the entrance to the structure. The gap voltage developed in this cavity ranges from about 80 to 130 kV with a 440-kV, 5-kA beam, so the four gaps can almost completely remove the energy from the properly phased portion of the beam. The gap spacing is tapered to maintain synchronism with the beam as it loses energy. However, efficiency is limited by transit time effects and

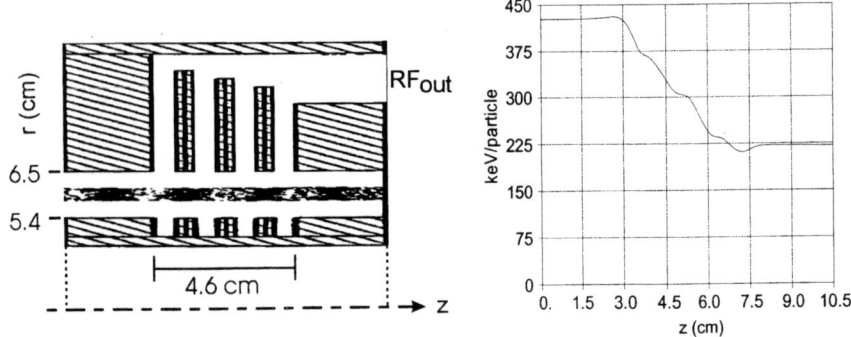

FIGURE 7. TKA output structure and simulated energy extraction.

by the fact that the out-of-phase portion of the beam is accelerated as it goes through the structure.

The average kinetic energy per particle (over all particles in the simulation) as a function of distance in the output structure is also shown in Fig. 7. This plot shows that most of the energy extraction occurs in the first three gaps. The 50% extraction efficiency corresponds to an output power of 1.1 GW. Still, with the multiple gap structure the peak surface field is comfortably below the breakdown threshold.

Ongoing work on the triaxial klystron is devoted to demonstrating the ability to achieve the predicted energy extraction using the same beam generator used for the L-band amplifier. Preliminary studies also indicate that two important upgrades are feasible. First, the triaxial geometry is compatible with PPM focusing, with magnets and pole pieces placed both inside and outside the beam radius (and outside the vacuum). Second, a thermionic gun can be implemented to allow repetitive operation.

ACKNOWLEDGMENTS

This work has been supported by the Air Force Research Laboratory (L-band klystron) and the Department of Energy (X-band klystron). The authors thank Mike Jabari and David van Doren for their diligent laboratory assistance.

REFERENCES

1. M. Friedman, J. Krall, Y.Y. Lau, and V. Serlin, "Efficient Generation of Multi-GW RF Power by a Klystron-like Amplifier," Rev. Sci. Instrum. **61**, 171 (1990);
2. M. Friedman, et al., "Efficient Conversion of the Energy of Intense Relativistic Electron Beams into RF," Phys. Rev. Lett. **75**, 1214 (1995).
3. M. Friedman, V. Serlin, M. Lampe, and R. Hubbard, "Intense Electron Beam Modulation by Inductively Loaded Wide Gaps for Relativistic Klystron Amplifiers," Phys. Rev. Lett. 74, 322 (1995).
4. B. Goplen, L. Ludeking, D. Smithe, and G. Warren, "User-configurable MAGIC for Electromagnetic PIC Calculations," Comp. Phys. Comm. 87, 54 (1995).
5. K.J. Hendricks, P.D. Coleman, R.W. Lemke, M.J. Arman, and L. Bowers, "Extraction of 1 GW of RF Power from an Injection Locked Relativistic Klystron Oscillator," Phys. Rev. Lett. **76**, 154 (1996).
6. V. Serlin and M. Friedman, "High-Frequency Operation of the Relativistic Klystron Amplifier," in *Intense Microwave and Particle Beams III*, H. Brandt, ed., SPIE Vol. 1629, pp. 8-13 (1992).

Klystron Life Results in Particle Accelerator Applications

Heinz Bohlen

CPI Wireless Solutions, Palo Alto, CA 94303, USA

Abstract. Based on reports contributed by various particle accelerator sites, among them DESY, CERN, and LANL, Weibull life time characteristics have been calculated for the klystrons used at these institutions. Supported by evaluations of the technologies and the operational conditions involved, the results, sometimes surprising and unexpected, present material that can be valuable for logistic considerations, the planning of future accelerators, and naturally for the design of future klystrons.

INTRODUCTION

Cost considerations play a predominant role in the planning phase of any major particle accelerator. When discussing the assumed operation expenses, the lifetime expectancy of the RF amplifier devices becomes an important issue. After listening to and participating in many debates on this subject, the author thought that a survey and analysis of actual lifetime results of klystrons used in major accelerators might provide a useful tool for pre-calculating the life expectancy and with that the costs for klystron replacement for any period of accelerator operation.

Data about the lifetime of failed klystrons and the running time of those in operation are meticulously collected and recorded at most accelerators. But they are difficult to interpret. Most klystron types do not behave like light bulbs, dying all after more or less the same operation time. In most cases klystrons are not made in very large numbers; their behavior differs from individual to individual. Beyond that, the manufacturers learn from failures during the initial time of the operation of the accelerator, which (hopefully) results in an improvement of the life expectancy for later klystrons. This altogether leads to statistics in which the average life of the failed klystrons is usually short and that of the survivors very uncertain.

This conflict can be solved by introducing the Cumulative Hazard Function $H(x)$ of the so-called Weibull distribution:

$$H(x) = x^\gamma$$

x is the lifetime of the individuals in this case, and γ is the shape parameter of the distribution. The Weibull distribution takes into account all members of a population, dead or alive. It does this by increasing the hazard rate accordingly each time an individual dies that is survived by younger members of the population. Since each individual's hazard rate simply equals its percentage of the total population, a cumulative hazard rate of 100 % denotes average lifetime. It has become custom to plot the logarithm of the hours (x) versus the logarithm of the Cumulative Hazard (H); this method leads to a linear dependency if the shape parameter γ remains constant over the life of the population.

For the reason of economy of space the following presentation is limited to only a few examples. It (yet) lacks a survey on high-power short-pulse klystrons. However, the author hopes that these examples are at some degree representative for a major range of klystron products, especially for the reason that klystrons of five major manufacturers are involved in the survey.

LONG-PULSE KLYSTRONS[1]

The data presented in this paragraph are lifetime results of klystrons VA-862A, operated in the LAMPF accelerator at Los Alamos National Laboratories. This is an extract of the klystron's specification:

Frequency	805	MHz
Peak output power	1.25	MW
Pulse length	1.1	ms
Duty cycle	13.2	%
Beam voltage	85	kV
Gain	60	dB
Efficiency	50	%

The klystron features a mod-anode, water-cooled body and collector, an oil-insulated gun and electromagnetic focusing. It is equipped with a fairly large oxide cathode with a surface of almost 120 cm^2; the resulting specific peak emission density is a comfortable 0.25 A/ cm^2. There are practically four different populations of this klystron type to be observed at LAMPF:

    I: new klystrons by manufacturer A (69 members, 38 survivors)
    II: new klystrons by manufacturer B (26 members, 2 survivors)
    III: type I repaired by LANL (31 members, 11 survivors)
    IV: type II repaired by LANL (29 members, 12 survivors),

and we will learn that their behavior is not equal.

---

[1] Data presented in this paragraph by courtesy of Los Alamos National Laboratories

Weibull Distributions

Our first example (Fig. 1) was the attempt to list populations I and II (all new klystrons) in one diagram. The curve clearly shows a change in the shape parameter (the klystron life seems to grow with increasing age).

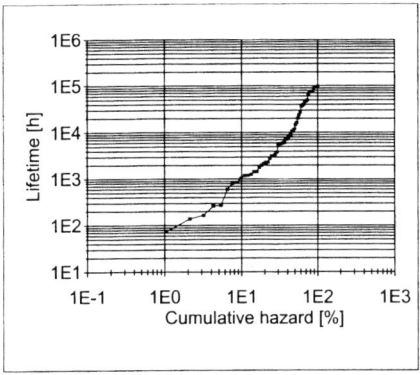

Figure 1. Weibull distribution of all new LAMPF klystrons (populations I and II)

In the next two figures (Fig. 2, Fig. 3) the two populations have been separated and now each of them shows a reasonable, consistent distribution.

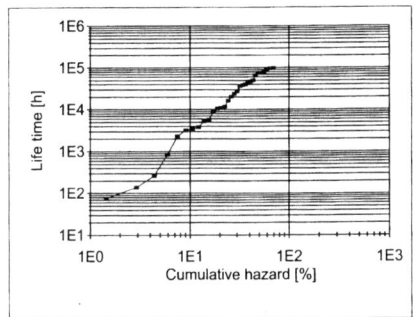 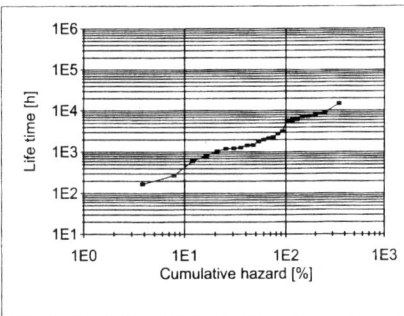

Figure 2. Distribution of population I   **Figure 3.** Distribution of population II

There are, however, considerable differences with regard to lifetime expectation. While population I, though marred by a number of early failures, aims at 200,000 hours average lifetime, population II reaches just a little more than 4,000 hours. This strong differentiation does not continue once the klystrons have been repaired at LANL. As Figures 4 and 5 demonstrate, repaired klystrons of any origin have an average lifetime expectancy of about 35,000 hours.

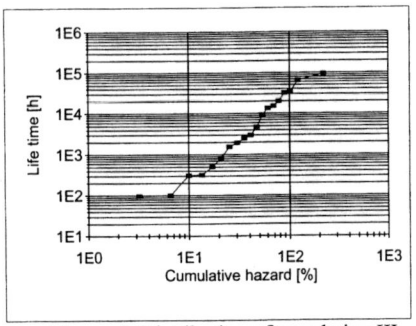
Figure 4. Distribution of population III

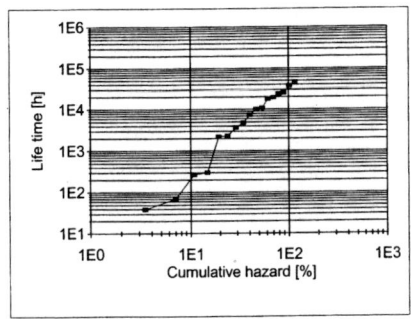
Figure 5. Distribution of population IV

Failure Modes and Causes

Despite the fact that electrical and mechanical layout of the klystrons in all four populations is identical, the pattern of failure modes differs from population to population as significant as the life time expectancy does:

Table 1. Leading failure modes (Long-pulse klystrons)

Population	Leading Failure Modes	Remarks
I	Vacuum failure; cavity 1 damage; open filament	Cavity 1 damage caused by solenoid problems
II	Arcing mod-anode to cathode; multipactoring ("glitches")	
III, IV	Vacuum failure; low emission; multipactoring ("glitches")	

This leads to the conclusion that a deciding factor for the lifetime expectancy of this group of klystrons is related to the manufacturer's processes. The striking differences in average life in this case are obviously caused by dissimilarities in handling and control of surface chemistry, brazing and welding, cathode coating and exhaust.

SUPER-POWER CW KLYSTRONS[2]

The following four populations have been chosen as typical for the CW class of super-power klystrons:

---

[2] Data presented in this paragraph by courtesy of CERN and DESY.

Table 2. Super-power CW klystrons in LEP and HERA accelerators

Population	Accelerator	Manufacturer	Members	Survivors
I	LEP	C	31	16
II	LEP	D	29	15
III	LEP	E	16	13
IV	HERA	C	86	23

These klystrons belong into the same power class and share the basic technology, but each population features an individual design. The main specification data are:

Population	I-III	IV	
Frequency	352.2	499.7	MHz
CW output power		1300 800	kW
Beam voltage	100	76	kV
Gain	41	42	dB
Efficiency	67	62	%

Weibull Distributions

Though the Weibull distribution characteristics of populations I (Fig. 6) and II (Fig. 7) arrive at similar lifetime expectancies, they are nevertheless very different. Population I (manufacturer C) suffers from early failures but finally reaches an average life of about 33,000 hours, while population II (manufacturer D) shows much higher consistency but slightly shorter average life (25,000 hours).

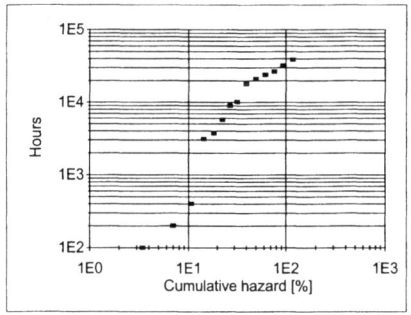
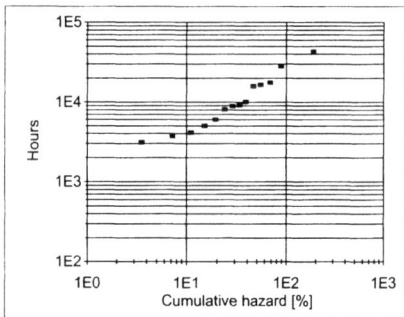

Fig. 6. Population I in LEP   **Fig. 7.** Population II in LEP

Population III lost only 3 members so far, which makes it difficult to plot a sensible distribution, but Fig. 8, presenting populations I through III combined, indicates that it fits very well into the same characteristic, predicting a combined average life of 29,000 hours for the mix of LEP klystrons.

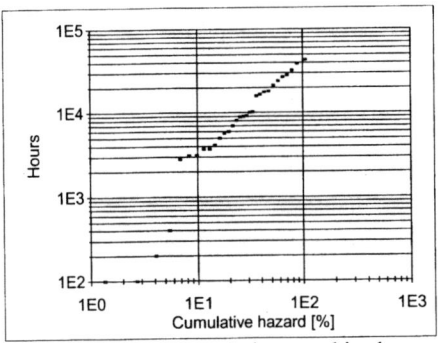

Fig. 8. All LEP populations combined

Population IV (Fig. 9) contains the first ever super-power klystrons that were manufactured in larger quantities. This may be the reason for their lower average life (23,000 hours). However, some of the original problems must have been mended, because the average operation time of the 23 survivors has reached 24,000 hours and thus is higher than the average life of the total population as calculated according to Weibull. DESY actually confirms that the last series of klystrons shows a tendency towards 35,000 hours of average lifetime.

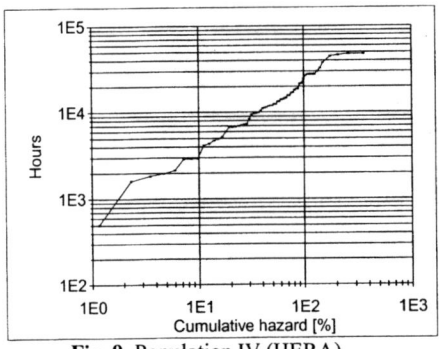

Fig. 9. Population IV (HERA)

Failure Modes and Causes

Discussions about high-efficiency CW klystrons among experts usually culminate in the topic of instabilities. The kind most dreaded by operators of particle accelerators are oscillations, also called spurious noise, that appear a few megahertz below and beyond the operational frequency. They are caused by backstreaming electrons when high output power combines with high efficiency. The LEP klystrons feature an efficiency around 67 %, which is high at any rate.

However, when inspecting the list of failure modes at LEP (Table 3), it becomes obvious that instabilities range at the end of the list. In other words: Spurious noise is not inevitably linked with operation at high output power and high efficiency; the manufacturers have means to suppress it.

Table 3. Failure modes of LEP klystrons

Failure mode	Number	Percentage
High voltage arcs in gun	11	36
Vacuum leaks	9	29
Emission loss	3	10
Arcing in RF output	2	6
Bucking coil short	2	6
Heater short	1	3
Instabilities	1	3

The obviously predominant failure mode, even more frequent than the finally inevitable vacuum leak, is arcing in the electron gun. The cause is known: barium evaporating from the cathode is deposited on high-voltage electrodes, preferably on the focus electrode and on the anode. The result is small-area emission from the focusing electrode and/or flaking material on focusing electrode and anode. Both lead to high-voltage arcing. It seems that cathode temperature reduction, which sharply reduces barium evaporation, is the most promising measure to prolong the average life of CW klystrons.

The ability of manufacturers to handle this tool obviously differs: while the main failure mode of klystrons produced by manufacturer C is indeed gun arcing, the klystrons of manufacturers D and E die preferably from vacuum leaks, without achieving higher average life, however.

CONCLUSION

It becomes obvious that high-power klystrons in particle accelerators can reach high average life results, long-pulse klystrons in the order of 100,000 hours or more, CW klystrons in excess of 25,000 hours. That certain populations do not arrive at these numbers doesn't seem to be caused by the electrical or mechanical design of the devices in most of the cases. The reason appears to be rather the differing expertise of manufacturers regarding essential processes, mostly related to cathode and vacuum surface technology.

ACKNOWLEDGEMENTS

The author feels indebted to the persons who provided the essential material for this paper: Michael Lynch of the Los Alamos National Laboratory, Hans Frischholz of CERN, and Michael Ebert of DESY.

Recent Progress in Multi-Beam Klystron in IECAS

Ding Yaogen

Institute of Electronics, Chinese Academy of Sciences
Beijing 100080, China

Abstract. Recent research progresses in MBK in IECAS are briefly introduced in the paper. The S-band MBKs of IECAS have peak power of 120-250 kW, average power of 4-9 kW, efficiency of 35-45 %, gain of 41-46 dB, beam voltage of 15-19 kV, and weight of 40-45 kg. Some key technical problems of MBK are also described and discussed. Among them, improving the design of MBK to obtain the required bandwidth, raising beam transmission to increase average power, eliminating oscillation and spray spectrum, overcoming window breakdown caused by magic mode, reducing breakdown times of electron gun, are most important things for the practical MBK. Besides, further research work in MBK in IECAS is commented.

INTRODUCTION

Multi-Beam Klystron (MBK) is a new type of high power broadband microwave vacuum devices which has the advantages of wide bandwidth on low power level (several tens kW to several hundreds kW), high average output power, high gain and high efficiency, low beam voltage, small dimension and light weight. These advantages have a strong appeal to microwave electronic system designers who are interest in developing the compact transmitter.

Institute of Electronics, Chinese Academy of Sciences (IECAS) began to develop MBKs at the beginning of 1990s. Since then many types of MBKs have been developed, and several of them were used in practice. The main research progress of MBK conducted in IECAS is briefly introduced in this paper. The key technical problems met in MBK are discussed, and further research work on MBK is also commented.

RECENT PROGRESS OF MBK IN IECAS

Recent years, four types of S-band MBKs have been developed in IECAS. The main specifications of these MBK are listed in Table 1. These MBK have peak power of 120-250 kW, average power of 4-9 kW, efficiency of 35-45%, gain of 41-46 dB, bandwidth of 7-9.8%, beam voltage of 15-19 kV. All of these MBK adopt Periodic Reversal Permanent Magnet (PRPM) focusing system. The weight of these tubes including focusing system is only 40-45 kg, and the length of these tubes is less then 800 mm. These MBK have DC beam transmission of 90-95%,

TABLE 1. The specifications of S-band MBK.

Type	P_{peak} (kW)	P_{ave} (kW)	η (%)	G (dB)	δf/f (%)	V_0 (kV)	I_0 (A)
KS-57	>205	>4.1	>36.9	>45.3	8.3	18.2	30.5
KS-60	>218	>8.2	>43.0	>46	7.0	18.3	27.7
KS-72	>175	>7.0	>36.8	>41.8	9.8	18.1	26.3
KS-93	>127	>6.4	>35.3	>44	8.8	15.2	23.6

and RF beam transmission of 60-85%. The DC beam transmission variation with beam voltage for KS-60 is shown on Figure 1. For beam voltage from 17-20 kV, DC beam transmission is 92-94%. The power–frequency characteristics of KS-60 2#E is given on Figure 2.

IECAS is developing L-, C- and X- band MBK with power level of 75-600 kW. The design and experimental parameters of L-band MBK KL-81 are listed on Table 2.

Comparing L-band MBK KL-81 with L-band TWT PT6049 made by TMD, KL-81 has low beam voltage (18-19 kV), and short tube length (850 mm). The PT6049 has beam voltage of 43 kV and tube length of 1920 mm. Besides, KL-81 uses PRPM focusing system, and PT6049 adopts electromagnet focusing system. The main technical problems met on KL-81 will be discussed in the next section.

Two types of C-band MBK are also developing in IECAS. The experimental tube of KC-79 has peak power of 200 kW, average power of 6 kW, bandwidth of 220 MHz, and efficiency of 30 %. Both electromagnet and PRPM focusing system are adopted in this tube. Another type of C-band MBK KC-92 with power level of 120 kW, and bandwidth of 7%, is developing.

TABLE 2. The main parameters of KL-81.

Parameters	DESIGN	EXPERIMENT
Operating frequency range	L-band	L-band
Bandwidth	11.4%	10.8%
Peak output power	180 kW	140 kW
Average output power	3.6 kW	3.0 kW
Efficiency	35 %	
Gain	41 dB	

KEY TECHNOLOGY PROBLEMS OF MBK

Several technical problems met in MBK are described and discussed in this section.

Design improvement by using 2.5 D klystron computer code Arsenal.MSU

Usually, 1D klystron computer code KLY6 is used to simulate beam-wave interaction of MBK in IECAS. There is good agreement between design and experiment parameters for S-band MBK. But there is discrepancy for C-band MBK, as shown in Figure 3. The zero point of gain and efficiency appears on high frequency band. The main reason is that 1D code can not consider the effect of beam ripple and beam interception. By using 2.5D code Arsenal, these phenomena can be explained. Through improving beam quality and increasing the number of bunching cavity, the zero point of gain and efficiency can be overcome.

Research on multi-beam electron gun and cathode with high emission current density

Main research work conducted in IECAS includes following aspects:

a) Increasing emission current density and lifetime of dispense cathode by improving the machining and technology of tungsten matrix, coating alloy film (Os-Hf, Os-W), and decreasing the brazing temperature for cathode and its supporter.

b) Reducing times of breakdown in electron gun region by reducing the evaporation of cathode and the temperature of controlling electrode, and coating Hf film on it to suppress thermal emission.

c) Reducing ready time of MBK by decreasing the relative expansion between cathode and controlling electrode.

Research on improving beam transmission

The beam transmission of 90-95% for DC State and 70-80% for RF State has been obtained for S-band MBK KS-60. But some types of MBKs have very lower RF beam transmission (lower than 60%). The low RF beam transmission has big effect on stability, output spectrum and average power. For example, owing to the beam interception in output region, the magnetic pole piece was melted during hot test for C-band MBK KC-79 with electromagnet focusing. The beam transmission can be improved through computer simulation and careful adjust of focusing system. The most important thing is reducing the transverse magnetic field for outer layer beams.

Oscillation and spray spectrum

There is feedback mechanism to generate oscillation and spray spectrum output circuit with double gap coupling cavity. In L-band MBK KL-81, an oscillation mode with frequency of 3947 MHz exists. The 3D electromagnetic field simulation shows that this mode has 2π mode field structure, and $\beta eP=430^0$ which

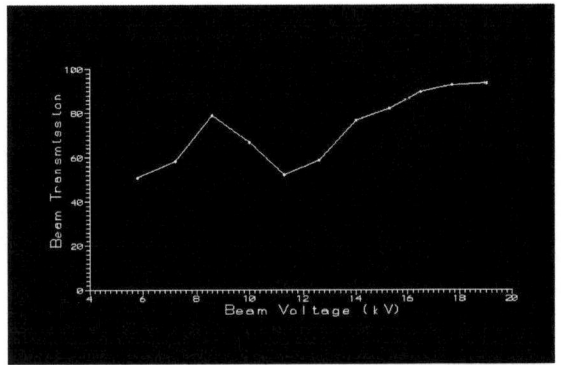

FIGURE 1. Beam transmission vs. beam voltage.

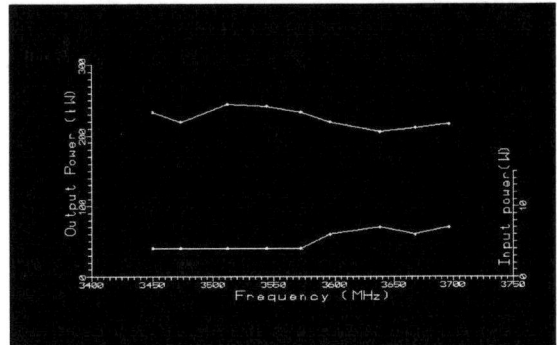

FIGURE 2. Power-frequency characteristics of KS 60 2#E.

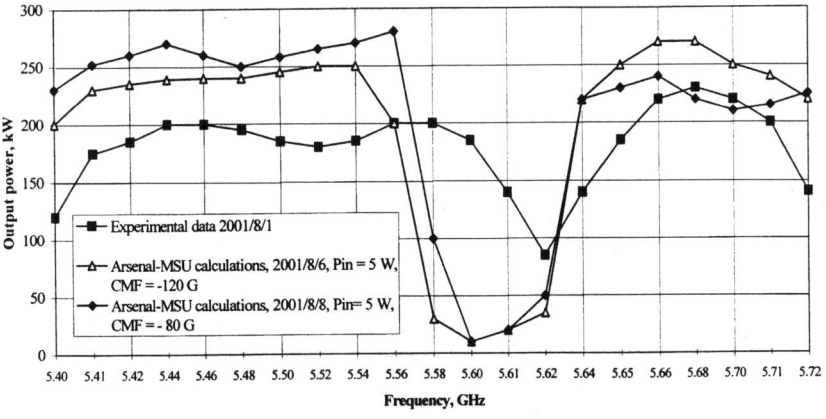

FIGURE 3. Power-frequency characteristics of KC-79 (calculation and experiment results)

is located in negative electron conductance region. By adding additional absorbing cavity, the oscillation was eliminated. Also, when there exists RF input signal, the spray spectrum of 1513 MHz appears. It is another 2π mode of output structure, which is near the operating π mode. This spray spectrum is generated by reflection or secondary electrons, and can be eliminated by improving RF beam transmission, and lowing Q of absorbing cavity. The spray spectrum can be also excited by the RF energy leakage from insulator of collector or by the reflecting and secondary electron. It is necessary to design the collector structure carefully to eliminate the spray spectrum. The RF energy leakage can be absorbed through cooling water.

The breakdown of high power output window

Several pieces of S-band MBK were broken owing to the failure of output window. Among them, two are caused by the magic mode of pillbox window, and the others are owing to the spark of output waveguide system. 3D simulation of pillbox window shows that the magic mode has TM_{01} field structure and can be excited by asymmetry of window structure. There is normal electric field component on the surface of ceramic for this mode, which will cause multipactor effect, generate abnormal heat loss, and destroy the window. By changing the diameter of circular waveguide, the magic mode can be moved out of operating frequency band. The new window can endure average power of more than 9 kW.

DISCUSSION AND CONCLUSION

Several types of MBK have been developed in IECAS recent years. The frequency range of MBK developed in IECAS occupies whole microwave frequency band from L-band to X-band. Main purpose of IECAS is to develop compact broadband MBK with peak power of several hundreds kW, average power of several kW to several tens kW (even several hundreds kW), bandwidth of 5-15%. The further research work on MBK includes:

1) Computer simulation of multi-beam electron gun, focusing system, RF system, beam-wave interaction, and heat dissipation by using 3D software

2) Improving the machining and technology of MBK to raise stability and lifetime

3) Developing new electrodynamic system to extend the bandwidth, increase output power in high frequency band

The author wishes to acknowledge the dedicated efforts of his colleagues and graduates for the development of MBKs.

REFERENCES

1. Ding, Yaogen "The state of art and trend of multi-beam klystron", 25th International conference on Infrared and Millimeter Wave, September 14-16, 2000, Beijing China.
2. Ding, Yaogen "Study on Electron Optics of Multi-Beam Klystron", Journal of Electronics (Chinese), Vol.22, No.3, pp485-491.

Inductive Output Tubes – Status and Future Direction[*]

Heinz Bohlen

CPI Wireless Solutions, Palo Alto, CA 94303, USA

Abstract. Invented in 1938, at the same time as the klystron, it took the Inductive Output Tube (IOT) more than 40 years to surface as a useful device. Its progress after that event was rapid. Though plagued by teething problems in the beginning, it has since replaced the klystron as a TV amplifier in UHF, and it holds its own against the solid-state competition in that application. The IOT also shows much promise as a high-power amplifier, but early developments in this direction have remained solitary events so far. The paper discusses the causes and the potential of the device for future high-power applications.

1. INTRODUCTION

The Inductive Output Tube (IOT) exists as a commercial product for almost two decades by now. But there is still a lot of confusion about this tube type. What is the difference between a Klystrode, an IOT, or a CEA? Is the IOT a linear beam tube like the klystron or the TWT, or is it just a special version of the good old tetrode? Is it useful as a TV amplifier only? What limits its output power and frequency range? This paper will try to give answers to questions of this kind.

2. THE IOT'S BACKGROUND, OR HOW TO NAME A DEVICE

2.1 Andrew Haeff's Invention

The IOT has a colorful history. In February 1939, in a period when the race was on for microwave power devices that could be used for radar, Andrew V. Haeff, at that time working for RCA in Harrison, N. J., published a paper about "an ultra-high-frequency power amplifier of novel design" [1]. It is the first IOT, but in this paper it has no name yet. It is merely described as a gridded device combined with a "tank circuit", a reentrant cavity that is obviously the main issue of the paper. Haeff

---

[*] This is a modified version of a paper by the same author presented at the "Displays and Vacuum Electronics" convention at Garmisch-Partenkirchen in May 2001.

produced 110 W of output power at 450 MHz with this amplifier, reaching an efficiency of 35 % with the aid of collector potential depression. In his next paper on the subject [2] about one year later, in March 1940, Haeff called his device an "inductive-output amplifier", referring to the way the output power is extracted from the electron beam. This time the emphasis, probably as an answer to the Varian brothers' publication on the klystron in May 1939, was on wide-band operation. Haeff reported 10 W of output power at 500 MHz, 10 MHz bandwidth, 10 dB gain and 25 % efficiency.

Let's have a look at the kind of device that Haeff had invented. The IOT principle is neither new nor complicated but relatively little known. Hence a short description might be useful. The easiest way to explain the basic device is by comparison with a conventional planar tetrode, the screen grid of which is on anode potential and is provided with a hole to let the electron beam pass (see Figure 1).

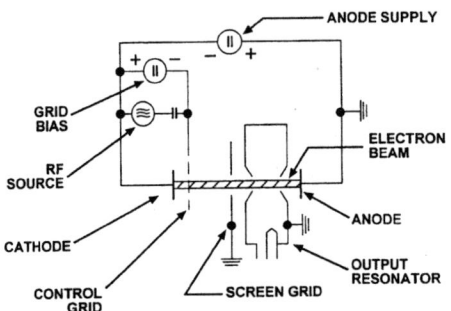

FIGURE 1. Inductive Output Tube (schematic)

Thus the electron beam, after being high-frequency modulated between cathode and control grid and accelerated between control and screen grids, enters a dc-free space behind the screen grid. There it passes through the interaction gap of a resonator that is tuned to the modulation frequency, converting a considerable part of its energy into rf output power, and is finally collected by the anode.

It is not immediately obvious what happened to the properties of the tetrode by this modification but the consequences, especially at frequencies beyond VHF, are enormous:

- The anode potential is not modulated by rf. This not only eliminates any need for neutralization, but practically also the transit time problem between control grid and anode, enabling the IOT to operate with higher efficiency than the tetrode at high frequencies.
- The same feature also removes the need to compromise between gain and efficiency by the choice of screen grid voltage; the screen grid potential now being equal to the anode potential endows the IOT with considerably more gain than the tetrode.
- The screen grid becomes a practically non-intercepting electrode. This removes a major obstacle for achieving high average power.

By these virtues the IOT becomes a competitor to the klystron in the UHF range, too, and in this comparison it can compensate its still much lower gain by the tetrode properties that it maintains:
- High linearity;
- High efficiency in class C operation;
- Slow drop of efficiency at reduced output power;
- Because the output power still increases at the point of maximum efficiency it can be controlled by feedback.

Though usually designed around a linear electron beam, the IOT is not a linear beam tube in the classical sense; any transit time effects, like bunching, are not essential for its basic performance. There have been, however, versions patented that intend to use transit time effects for a further increase of efficiency. In a nutshell: The IOT is a special tetrode with considerably improved properties especially in the UHF range.

One should have guessed that in the light of all these merits military experts would have jumped at this invention. The opposite was the case. World War II began. Magnetrons and reflex-klystrons did not require any drive power, and thus there was no demand for the IOT. Five years after the war Haeff became the first recipient of the Harry Diamond award "for his contribution to the study of the interaction of electrons and radiation, and for his contribution to the storage tube art". But his IOT was forgotten.

2.2 The Rediscovery

In the early 1970s the energy crisis erupted. Trying to stem the tide, engineers concerned with high-power microwave devices started thinking harder about ways to increase efficiency. At Philips-Valvo in Hamburg this resulted in 1974 in the idea of a device that was considered a hybrid between a tetrode and a klystron. Consequently it was named "klystrode" [3]. Figure 2 shows a sketch that was made at that time. It is a tube in "modern" metal-ceramic technology, and the sketch clearly shows a "depressed collector". The output circuit was meant to be an external, detachable cavity; it is not shown here. A patent application for the klystrode was not successful, however, and a closer look at Figure 2 reveals why: it is an IOT, a device that the Valvo engineers had never before heard of. Anyway, Valvo claimed a utility model for the device, but decided not to produce the tube at that time, and the IOT dropped back into obscurity.

Thus it was up to another company, Varian/Eimac in San Carlos, California, to reinvent it. In 1982 this company independently arrived at the same technical solution, and also the same name for the device: klystrode [4]. Again the engineers learned about Haeff's achievement only after the fact.

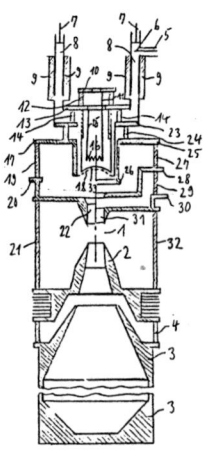

FIGURE 2. "Klystrode" sketch (1974)

This time the device, X-2250, went into production as a high-power (up to 28 kW) TV-sound amplifier. The output circuit was still narrow-band; thus vision amplification was not possible yet. The efficiency reached 58 %, and the gain was about 21 dB at 775 MHz. Figure 3 shows the outline of the tube.

Two years later a modified amplifier, X-2251, equipped with a secondary output circuit to increase bandwidth, was ready for 32 kW of vision output power. The following introduction period of the new technology was not easy and required much pioneering spirit. Teething problems, mainly grid emission, caused considerable drawbacks, but by about 1990 the TV transmitter market demand for the new device had grown large enough that other manufacturers were tempted to participate.

In April 1990, EEV in Chelmsford, UK, announced its 70 kW IOT7360 [5]. Since in 1987 the expression "Klystrode®" had become a trademark of Varian-Eimac, EEV decided to revert to Haeff's old name for the device, and thus the generic expression

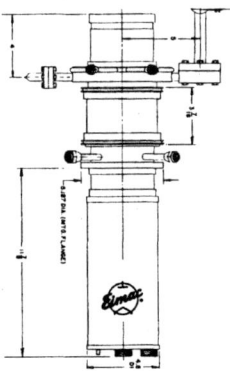

FIGURE 3. X-2250, the first modern IOT (1982)

became "IOT". Not only did the tube set a new technical standard, but in addition the arrival of a second manufacturer signaled to the users that IOT technology finally was here to stay.

Meanwhile the pioneers at Varian-Eimac were busy promoting the new device into the realm of higher power. With the support of Los Alamos National Laboratory, the high-power long-pulse IOT X-2259 was developed in the late 1980s. It reached 500 kW of peak power and 70 % efficiency at 425 MHz [6]. It was followed by the first IOT to develop high continuous-wave power, the 2KDW250PA, known as the "Chalk River Tube" (Figure 4) and producing 250 kW CW output power and around 73 % efficiency at 267 MHz [7].

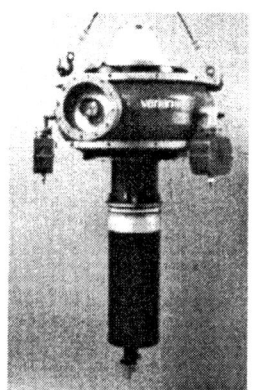

FIGURE 4. 250 kW CW IOT

In just a decade the IOT had grown from an idea into a high-power device. Understandably, not all technical issues had found a satisfactory solution in that short time interval.

3. STATUS

3.1 The TV IOT, or How Many Collectors Do We Need

The final 10 years of the 20th century brought about a host of new types of TV IOTs and a rapid increase in operational sockets. Philips-Valvo reconsidered its position and produced an IOT, and companies like Thomson, Litton and Istok joined in. Istok even developed a multi-beam version. TV-IOTs are presently operated in about 1200 sockets worldwide.

High efficiency, even at reduced signal levels, and high linearity make the IOT a superior device for high-power terrestrial television transmitters, and an extremely economical competitor to solid-state transmitters even down to output power levels as low as 10 kW. In analog TV operation, its linearity properties permit common amplification of vision and sound signal up to a nominal system output power of 70 kW, corresponding to 77 kW vision (peak sync) plus 7.7 kW sound power at the output flange of the cavity. Correspondingly, 150 kW of peak output power are reached in digital TV operation. Devices with higher output power would be feasible but are not required for present day TV transmission.

High basic efficiency and only a moderate loss of efficiency at reduced signal level are an intrinsic IOT property. After all, it is a tetrode. But since its collector (unlike the anode in a traditional tetrode) is separated from the rf circuit, single stage or even multistage collector depression (MSDC) is applicable to an IOT, aiding further improvement of its efficiency. The graph in Figure 5 [8] presents the result of applying single stage collector depression to a broadband 60 kW IOT (CPI-Eimac K3D60). This single stage depression requires three isolated electrodes, but only one additional power supply. One of the two remaining electrodes is connected to ground, the other to cathode potential.

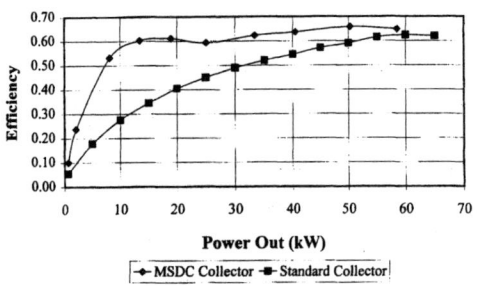

FIGURE 5. Efficiency improvement by single stage collector depression

The designers of the Litton CEA (Constant Efficiency Amplifier) went one or rather two steps further [9]. Their device uses three additional power supplies and a total of five isolated electrodes. The test results differ from those shown in Fig. 5 only marginally. It becomes evident that there is little room for further improvement, and that for more depression stages the balance between benefit and growing complexity has to be carefully contemplated.

Despite the multitude of types that has been developed by different manufacturers, variety can actually be found in detail only. The TV application field is very much user-defined. Hence issues like reliability, stability of operation, simplicity and ease of handling score highest on the priority list; efficiency and gain are important but ranked lower. Thus presently water-cooling is preferred to air cooling, non-depressed operation to collector depression, drop-in replacement to traditional assembly. Due to the requirement of tunability over the complete UHF-TV band, all existing TV-IOTs use external cavities.

The K2-series (CPI-Eimac) is a typical example for this technology. The IOT itself is available in four different power capabilities from 40 kW to 110 kW (K2D40W... K2D110W) but always in the same embodiment. Thus one set of hardware assembly (cavities, focus set, connectors, cooling jacket, etc.) covers the entire power and frequency range.

The IOT itself (Figure 6) is of the "drop-in" replacement type, permitting exchange of tubes within minutes. Visible are (from top) cathode contact ring, high-voltage insulator, anode, output cavity ceramic window, collector. Input cavity window, grid contact and heater contact are hidden inside the cathode contact ring where an ion getter pump is partly visible.

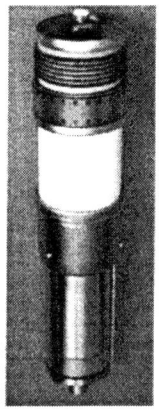

FIGURE 6. K2-series IOT

The picture of the hardware assembly (Figure 7) shows (from top) the input cavity, magnet mount containing the output circuit set and the support cart with collector water jacket. The output coupling unit is pointing upward from the output cavity.

FIGURE 7. K2-series assembly

3.2 The HOM-IOT

Scientific applications such as proton accelerators require UHF amplifiers with output power capability in the MW range. Super power klystrons are available for this service. However, IOTs offer some advantages over klystrons in terms of efficiency:
- Higher basic efficiency;
- Power regulation via feedback possible at full output power (no saturation);
- Inconsequential drop of efficiency at reduced output power.

The Higher Order Mode IOT (HOM-IOT) is an IOT version that employs cavities of the TM_{0n0} type (n > 1). Its cathode/grid system is arranged in an annular fashion, producing a cylindrical sheet beam that interacts with the outer electric field maximum in the cavities. This method permits the use of relatively large cathode areas and therefore possesses the added advantage of lower operational voltage and smaller size. Table 1 compares the (computed) data of an HOM-IOT and a klystron, both in 1 MW CW / 700 MHz proton accelerator service:

TABLE 1. HOM-IOT versus klystron: Comparison of computed data

	HOM-IOT	Klystron
Effective efficiency	73 %	60 %
Rel. power consumption	82 %	100 %
Assembly volume (approx.)	~ 1 m³	5 – 6 m³
Assembly weight (approx.)	~ 500 kg	2000 – 2500 kg
DC voltage	45 kV	90 kV

Based on these expectations, the development of a 700 MHz HOM-IOT, sponsored by the Los Alamos National Laboratory, has been carried out by CPI. The device employs the TM_{020} mode for both the input and the output cavity. The project aimed at creating an alternative power source for the APT proton accelerator. When the APT project was abandoned in 1999 in favor of a nuclear reactor solution, the HOM-IOT development was stopped.

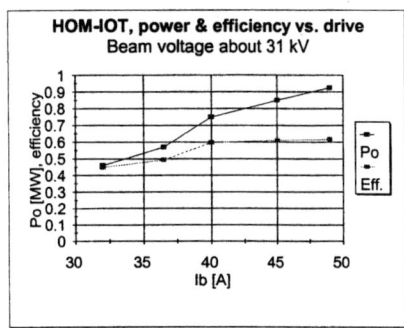

FIGURE 8. HOM-IOT, output power and efficiency versus drive (beam current)

At that time technological difficulties, like brazing a complex body of large diameter, had precluded a bake-out of the first sample at full temperature and thus prevented testing of the device in CW operation. However, the obtainable test data in short-pulse operation confirmed the results of the modeling effort to a high degree [10]. Operating at only 31 kV, the device produced an output power of 920 kW accompanied by an efficiency of 62 %, as shown in Figure 8.

Likewise confirmed was the expectation that the efficiency would drop only slowly at lower output power levels (Figure 9). Although not completed, the project supports the expectation that "low-voltage" devices with simultaneously high power and high efficiency can indeed be developed.

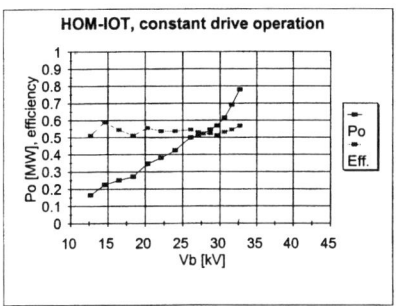

FIGURE 9. HOM-IOT, output power and efficiency versus beam voltage

4. FUTURE DIRECTION

The ability of the IOT to combine high linearity, high power output and high efficiency is likely to increase its field of applications further. Another very important feature is the possibility to pulse the device via its grid. Like with any tetrode or triode, this can be achieved by either pulsing the grid bias voltage, or the rf voltage. Particle accelerators and industrial heating processes will with some certainty make use of these properties. Several prototype transmitters have been installed recently. It was also found that the inter-pulse noise generated by an IOT is much lower than anticipated, suitable grid bias provided. Hence radar application add a further facet to the application range of IOTs. IOT development is expected to respond to these requirements with a variety of modifications, including other "low-voltage" devices with different shapes of electron beams, like for instance a coaxial IOT.

REFERENCES

1. Haeff, Andrew V.: An Ultra-High-Frequency Power Amplifier of Novel Design. Electronics, Feb. 1939, pp. 30 – 32

2. Haeff, Andrew V. and Leon S. Neergard: A Wide-Band Inductive-Output Amplifier. Proc. I.R.E., March 1940, pp.126 – 130
3. Bohlen, H. and H. Seifert: HF-Leistungssenderöhre mit hohem Wirkungsgrad im Mikrowellenbereich (Kurztitel: "Klystrode"), Nov. 1975. Patent Application P 25 46 358, Utility Model G 75 32 912.3
4. Preist, D. H. and M. B. Shrader: The Klystrode – An Unusual Transmitting Tube with Potential for UHF-TV. Proc. IEEE, Vol. 70, No. 11, Nov. 1982. pp. 1318 – 1325
5. Bohlen, H., G. T. Clayworth, R. Heppinstall, and D. M. Wilcox: Improved Technological Solutions for UHF Power Tubes. NAB Convention, April 1990
6. Preist, D. H. and M. B. Shrader: A High-Power Klystrode® with Potential for Space Application. IEEE Trans. on Electron Devices, Vol. 38, No. 10, Oct. 1991, pp. 2205 – 2211
7. Shrader, M. B., D. H. Preist and R. N. Tornoe: The 267 MHz High Power CW Klystrode® Amplifier. International J. of High Speed Electronics and Systems, Vol. 4, No. 4, Dec. 1993
8. Yates, C., Y. Li and E. McCune: Performance Characteristics of An MSDC IOT Amplifier. IEEE Trans. on Electron Devices, Vol. 48, No. 1, pp. 116 – 121
9. Symons, R. et al.: Prototype Constant Efficiency Amplifiers. IEEE Trans. on Broadcasting, Vol. 47, No. 2, June 2001
10. Bohlen, H., E. Lien et al.: 700 MHz HOM-IOT: Design and First Test Results. International Vacuum Electronics Conference, Monterey, May 2000

Recent Progress in Understanding the Physics of Plasma-Filled, High-Power Microwave Sources

G. S. Nusinovich[‡], Y. P. Bliokh[§], T. M. Abu-Elfadl[‡], A. G. Shkvarunets[‡], D. M. Goebel[*], Y. Carmel[‡], T. M. Antonsen, Jr.[‡], and V. L. Granatstein[‡]

[‡]*IREAP, University of Maryland, College Park, MD, 20742-3511, USA*
[§]*Technion, Haifa, 32000 Israel*
[*]*Boeing EDD Inc., 3100 w. Lomita Blvd., Torrance, CA, 90505, USA*

Abstract. The use of plasmas for generating high-power microwaves is studied for more than 50 years. During the 1990's Plasma-Assisted Slow-wave Oscillators (PASOTRONs) were invented and actively developed at Hughes Research Lab (HRL). These devices have a number of unique and attractive features. However, the experiments at HRL showed that to explore these features a better understanding of the physics is necessary. The present paper is focused on the recent studies of various physical issues, which are important for the pasotron operation. This theoretical and experimental activity resulted in more than doubling the pasotron efficiency (from about 20% to more than 50%) in the experiments carried out at the University of Maryland.

INTRODUCTION

Here we will not attempt to describe a long history of the development of plasma-filled microwave sources, which was recently overviewed in Ref. 1. Let us mention only that during the 1990's two new promising concepts of plasma-filled microwave devices were suggested and successfully realized. The first concept is based on the use of hybrid modes [2], which can be formed in plasma-filled slow-wave structures (SWS) by superposition of the plasma waves and slow waves of the SWS. These two sorts of waves become coupled when their phase velocities are synchronized and also their transverse structures overlap. The use of this concept has led to substantial increase in the gain and bandwidth of coupled-cavity traveling-wave tubes [3].

The second concept was suggested and pursued at Hughes Research Lab, where the devices named PASOTRONs (this acronym stands for Plasma-Assisted Slow-wave Oscillators) were invented [4] and successfully developed [5]. Pasotrons are unique devices, in which the electron beam propagation through the interaction space is provided not by solenoids or magnets, as in conventional microwave tubes, but by ions, which compensate for the radial beam space charge force, and thus, cause the effect known as a Bennett pinch [6]. The absence of the guiding magnetic field makes a number of interesting effects possible in PASOTRONs. Just these effects, which were studied recently at the University of Maryland, will be considered below.

NON-STATIONARY PHENOMENA IN THE ELECTRON BEAM TRANSPORT [7]

Initially, an electron beam propagates in a neutral gas, where it diverges radially due to a non-compensated electric space charge field. The beam impact ionization creates a plasma in this gas. Then, the beam expels plasma electrons while the ions neutralize the beam space charge field, thus causing the beam pinching. Clearly, this process, at least in its initial stage, is an inherently non-stationary process. Moreover, this process is a self-consistent process, because the gas ionization depends on the radial profile of the beam, while, in turn, the beam profile depends on the presence of ions.

In principle, the ions oscillate in both the radial and axial directions in the potential well formed by the beam space charge field, when this field is not fully compensated. The estimates show that for typical parameters of PASOTRON's the frequency of ion transverse oscillations is on the order of 1 MHz (we assume that the volume is filled with the helium only), while the frequency of ion axial oscillations is much lower and also the gas ionization proceeds in a much slower time scale (on the order of 10 :sec). Therefore, one can treat ion transverse oscillations as relatively fast, and therefore, consider a slow evolution of the beam envelope in the presence of ions with the stationary transverse profile of the density. This evolution, in turn, depends on the relation between the ionization time and the period of ion axial oscillations.

A. Fast Ionization

When the ionization time is much smaller than the period of ion axial oscillations the ion axial motion can be neglected. (Indeed, in such a case the accumulation of ions destroys the potential well faster than the ions pass from one of its walls to another.) In this case the ion density in a given cross-section of the device is determined by the local ionization rate, which is important when the initial gas density distribution along the device axis is nonuniform. As a result, the beam envelope equation can be written as

$$\frac{d^2\rho}{d\xi^2} = \frac{F(k\tau)}{\rho}[1-\gamma^2(\eta+C)\tau]+\frac{T}{\rho^3}. \tag{1}$$

Here a number of normalized parameters is introduced: $\rho = a(z)/a(0)$ is the beam envelope radius $a(z)$ normalized to its value at the entrance, $\xi = \sqrt{2I_b/I_A}(1/\gamma\beta)[z/a(0)]$ is the normalized axial coordinate, I_b/I_A is the ratio of the beam current to the Alfven current, $I_A = (mc^3/e)\gamma\beta$, γ is the electron Lorentz factor, $\gamma = 1/\sqrt{1-\beta^2}$ and β is the initial axial velocity of electrons normalized to the speed of light. Also, in Eq. (1) the function $\eta = n_0(\xi)/n_0(0)$ describes the initial axial profile of the gas density for the gas (He) leaking from the gas-filled, plasma electron gun and $C = (\sigma_b/\sigma_{He})[n_{0,b}/n_{0,He}(0)]$ is the coefficient, which takes into account the presence of an additional background gas (Ar or Xe), (here σ's are corresponding ionization cross-sections and the index "b" designates the background gas). The function $F(k\tau)$ models the beam current ramp-up during the time t_{rise}, $I_b = F(t)I_{b0}$, with $F(t) = t/t_{rise}$ for $t \leq t_{rise}$ and $F(t) = 1$ otherwise. Here $\tau = t/t_{ion}(0)$ is the time normalized to the ionization time for the gas density at the entrance, $t_{ion}(0) = 1/\sigma v_z n_0(0)$, and correspondingly, $k = t_{ion}(0)/t_{rise}$. Finally, the last term in the right-hand side of Eq. (1) describes the effect of a spread in transverse velocities. Here, parameter $T = \gamma^2 \beta^2 (I_A/2I_b) < \alpha_0^2 >$, where $\alpha_0 = \beta_{\perp 0}/\beta_{z0}$ is the transverse-to-axial velocity ratio at the entrance, is proportional to ε^2, where ε is the beam emittance.

This equation was studied for the exponentially decaying gas density profile, $\eta(\xi) = \exp(-\xi/L)$ in the presence or absence of the background gas and in the presence or absence of the initial transverse velocities of electrons. In a beam with a finite emittance, the stationary profile of the envelope, as is shown in Fig. 1, exhibits axial pulsations. The beam reaches this stationary state in a time scale on the order of the ionization time for the gas density near the entrance. Note that the position of the first focal plane shown in Fig. 1 agrees well with the results of the analysis of the stationary beam envelope equation presented in Ref. 8.

B. Slow Ionization

In the limiting case of a slow ionization $t_{ion} \gg t_{\parallel}$ the ions make a large number of axial oscillations during the ionization time. So, not only transverse but also axial profiles of the ion density become stationary during the ionization processes. When the gas density profile is practically constant, the ion axial motion does not play a big role in the beam pinching, as is shown in Fig. 2a, where the "fast" and "slow" ionization results are shown by the solid and dashed lines, respectively, for $L = 50$. Here (a), (b), (c) and (d) show the temporal beam radius evolution at different cross-sections of $\xi = 2, 4, 6$ and 8, respectively. On the contrary, in the case of gas localiza-

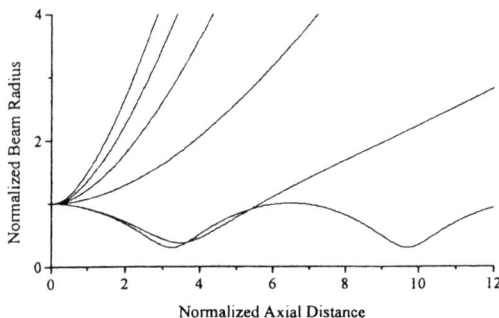

FIGURE 1. Onset of a stationary profile of the beam with a finite "temperature," T=0.05, in the absence of an additional background gas.

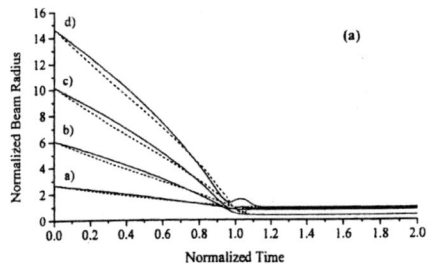

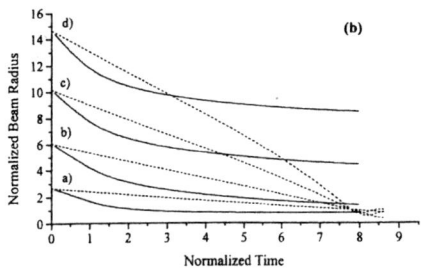

FIGURE 2. Beam pinching in the cases of fast (solid lines) and slow (dashed lines) ionizations: (a) L=50, (b) L=1.0. The cases a, b, c and d in each figure correspond to the cross-sections located in $\xi = 2, 4, 6$ and 8, respectively,

tion near the entrance, which is shown in Fig. 2b for $L = 1.0$, this axial ion motion is important, because it results in much faster beam pinching.

Also an intermediate case, $t_{ion} \sim t_{//}$ when the ions are described by kinetic equation for the ion distribution function $F = F(v_i)$,

$$\frac{\partial F}{\partial t} + \frac{\partial}{\partial z}(Fv_i) - \frac{e}{M}\frac{\partial \phi_0}{\partial z}\frac{\partial F}{\partial v_i} = S(z,t), \qquad (2)$$

has been studied. Here ϕ_0 is the potential describing the profile of the potential well responsible for ion axial oscillations and S is the ion source term, which is associated with the beam impact ionization, i.e., it depends on the electron beam parameters. Of course, the beam potential, in turn, depends on the ion density. Some results of the study of such a self-consistent problem are shown in Fig. 3, where the ion density per unit length, $N_i = \int_{-\infty}^{\infty} F dv_i$, is shown as the function of the normalized axial coordinate, ξ, and time, τ. The filaments shown in Fig. 3 can be interpreted as the axial acceleration of ions.

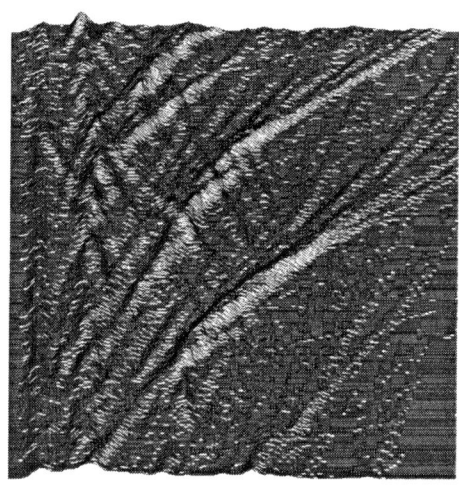

FIGURE 3. The ion density per unit length as the function of the normalized axial coordinate (horizontal axis) and normalized time (vertical axis).

ELECTRON 3-D MOTION IN THE INTERACTION SPACE

The ion focusing of an electron beam, which was discussed above, takes place in the absence of RF generation. However, since there is no strong guiding external magnetic field, in the presence of RF, the radial electric field of the wave may initiate the radial motion of electrons. This radial motion may enhance the wave excitation, because in the source term $<\vec{j}\vec{E}^*>$ describing the wave excitation (here angular brackets denote the averaging over the cross-section of the device), in addition to the term $j_z E_z^*$, which is the standard source term for the case of 1D-motion of electrons, also the term $\vec{j}_\perp \vec{E}_\perp^*$ appears. In particular, this term, as it was mentioned in Ref. 9, can be responsible for experimentally observed excitation of TE-waves [10].

The dispersion equation for such TWTs and BWOs, in which electrons can exhibit a 3-D motion, was derived a long time ago by Pierce [11]. In the variables normalized to the Pierce gain parameter C this equation can be written as

$$\gamma^2(\gamma-\delta)+1+q=0, \qquad (3)$$

where γ is the wave propagation constant, δ is the detuning between the wave phase velocity and initial electron velocity and q is in additional term responsible for the transverse interaction. It is obvious from Eq. (3), that this term increases the wave growth rate, and hence, can either increase the total gain in the device of a given length, or, alternatively, allows one to shorten the interaction length while realizing the same gain. This is the result of the small-signal analysis.

The large-signal theory was developed, in general, in Ref. 9, where it was shown that the radial motion of electrons at some detunings δ leads to the beam interception by a slow wave structure just at the instant when the field amplitude reaches its maximum. At the same time, at larger detunings the interception occurs later, which makes it possible to realize a high-efficiency operation without beam interception.

The efficiency enhancement was studied later in more detail in two subsequent papers [12,13]. In Ref. 12 the effect of the additional weak external magnetic field was analyzed. The theoretical analysis showed that an addition of a weak external magnetic field allows one to simultaneously enhance the interaction efficiency and avoid the beam interception by a slow-wave structure. For the parameters of the experimental PASOTRON-BWO which is currently under study at the University of Maryland, a corresponding magnetic field should not exceed 100G. This theoretical prediction was checked in the experiment [12], where it was shown that adding a 50G magnetic field causes the efficiency enhancement from 30% (at zero magnetic field) to 37%.

A more detailed analysis of the Helix PASOTRON-BWO efficiency was carried out in Ref. 13. The emphasis of this study was made on the role of the radial motion for electron coupling to the wave. Before presenting the results of this study, recall that the beam-wave interaction can be efficient when electrons are initially modulated by a weak RF field, and then electron bunches are decelerated by a strong field. In the BWO with a 1-D electron motion and a well-matched output, the situation is quite opposite: the RF field has the maximum at the beam entrance and equals zero at the beam exit. Therefore, a typical interaction efficiency in BWOs ranges from 10% to 20%.

The radial motion of electrons can drastically change this situation, if electrons are initially injected near the axis of a tube. Recall that the field of any slow wave is localized near the SWS and exponentially decays with the departure from it. So, when electrons are initially injected near the device axis, they start interacting with the weak field. Then, moving radially under the action of the radial component of the electric field of the wave, they start experiencing the strong RF field, which causes their deceleration. So, the effective amplitude of the RF field acting upon electrons moving both axially and radially may have an axial profile much more favorable for efficient interaction than in the case of a 1-D motion.

These simple arguments have been confirmed in the simulations presented in Ref. 13. Some results of these simulations are shown in Fig. 4. Here trajectories of two groups of electrons are shown: those injected near the axis are well bunched and move radially without substantial radial spread. On the contrary, the electrons having a large initial radius (1.5 cm) experience a large spread of trajectories. Another example is shown in Fig. 5 where the efficiency of electrons initially injected inside and outside of the helix are shown. As one can see, the efficiency of the first group exceeds the efficiency of the second group by more than two times. These preliminary steps resulted in simulations of the efficiency at the 55% level for the helix PASOTRON-BWO with an initially small beam radius. A corresponding dependence of the efficiency on the axial coordinate is shown in Fig. 6.

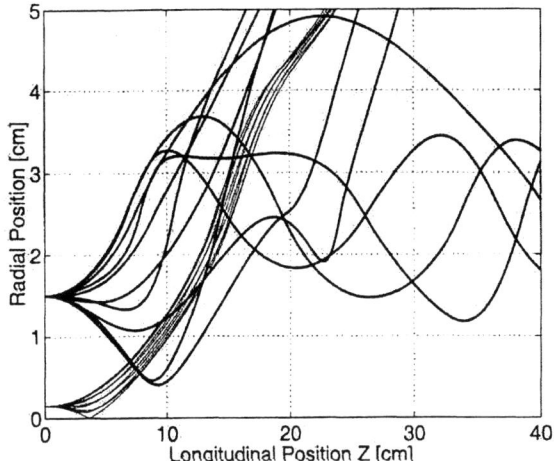

FIGURE 4. The radial trajectories of electrons with the different rays corresponding to different entrance phases for some selected beamlets. The operating frequency is $f=1.26$ GHz, beam current $I_b = 21$ Amp, beam voltage is 40 kV, and initial beam radius is 1.5 cm. The power in the backward direction at the beam entrance is about 1.5 MW. Electron beamlet entered close to the axis experiences small radial spread, while that entered away from the axis experiences large spread.

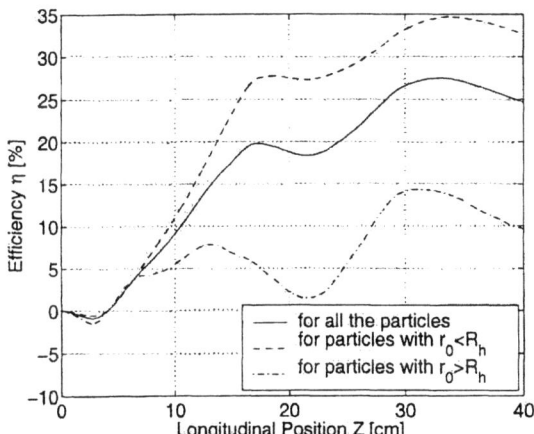

FIGURE 5. Electrons efficiency depends on their initial radial position in the beam. Electrons inside the helix have higher efficiency than those outside the helix.

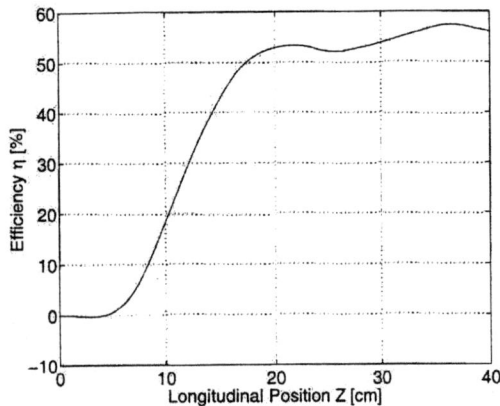

FIGURE 6. Electron efficiency of a beam with a small initial radius ($R_b = 1.5$ cm). Parameters are the same as in Fig. 4.

In accordance with these theoretical predictions some modifications in the PASOTRON tube were made. First, the plasma gun grid diameter was reduced from 8 cm to 4 cm, thus reducing the initial beam radius. Also, an adjustable upstream reflector was added to the waveguide arm leading to a matching load for controlling the Q-factors of competing axial modes. (Note that some issues in the competition of these modes were analyzed in Ref. 14.) So, the Q-factor of the desired mode was optimized for efficiency operation, while Q-factors of parasitic modes were simultaneously reduced.

These two modifications resulted in more than 50% efficiency operation at the 0.5 MW power level. This result is illustrated by Fig. 7, where the efficiency and power are shown as the functions of the cathode current. This experiment is described in more detail in Ref. 15.

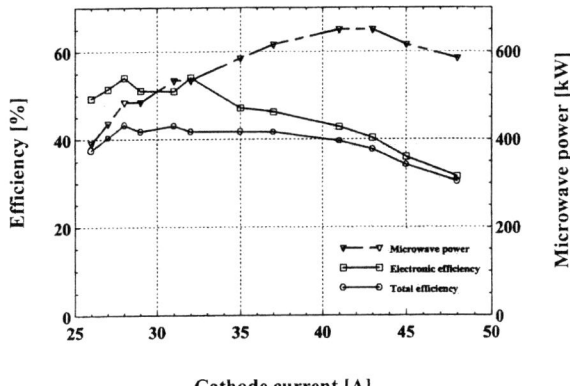

FIGURE 7. Pasotron power and efficiency as functions of the cathode current.

ACKNOWLEDGMENTS

This work has been supported by the Air Force Office for Scientific Research (New World Vistas Program).

REFERENCES

1. G. S. Nusinovich, Y. Carmel, T. M. Antonsen, Jr., D. M. Goebel, and J. Santoru, IEEE-PS **26**, 628 (1998).
2. N. I. Karbushev, Y. A. Kolosov, E. I. Ostrensky, and A. I. Polovkov, AIP Conf. Proc. 337, "Pulsed RF Sources for Linear Colliders," Montauk, NY, Oct. 1994, Ed. R. C. Fernow, AIP, New York, 1995, p. 360.
3. M. A. Zavjalov, L. A. Mitin, V. I. Perevodchikov, V. N. Tskhai, and A. L. Shapiro, IEEE-PS **22**, 600 (1994).
4. R. W. Schumacher et al., U.S. Patent #4912367, March 1990.
5. D. M. Goebel, J. M. Butler, R. W. Schumacher, J. Santoru, and R. L. Eisenhart, IEEE-PS **22**, 547 (1994)
6. W. H. Bennett, Phys. Rev. **45**, 890 (1934).
7. Yu. P. Bliokh and G. S. Nusinovich, IEEE-PS **29**, 951 (2001).
8. J. D. Lawson, "The Physics of Charged-Particle Beams," Oxford, U.K.: Oxford Univ. Press, 1977, Ch. IV.
9. G. S. Nusinovich and Yu. P. Bliokh, Phys. Rev. E **62**, 2657-2666 (2000).
10. D. M. Goebel, R. W. Schumacher, and R. L. Eisenhart, IEEE-PS **26**, 354-365 (1998).
11. J. R. Pierce, Traveling-Wave Tubes, Van Nostrand, Toronto, 1950, p. 173.
12. T. Abu-elfadl et al., Phys. Rev. E **63**, paper 066501 (June 2001).
13. T. Abu-elfadl et al., "Theory of Helix PASOTRON Backward Wave Oscillator," submitted to the Special Issue of IEEE-PS on HPM Generation.
14. G. S. Nusinovich and Yu. P. Bliokh, Phys. Plasmas **7**, 1294-1301 (2000).
15. A. G. Shkvarunets et al., "Realization of high efficiency in a plasma-assisted microwave source with two-dimensional electron motion," submitted to Phys. Rev. Lett.

Overview of LIGA Microfabrication

Jill Hruby

Sandia National Laboratories, Livermore, CA 94550 USA

Abstract. This paper is an overview of the LIGA technique, an increasingly accepted approach for fabricating metal, ceramic or plastic microdevices. The LIGA technique was invented in Germany in the early 1980s and the acronym derives from the words LIthographie, Galvanoformung, Abformung meaning Lithography, Electroplating, and Molding in English. The paper is presented as an abbreviated set of annotated overheads used for the conference presentation and some summary remarks.

SUMMARY OF LIGA PROCESS

Baseline LIGA Process

The LIGA process uses x-ray lithography accomplished using synchrotron radiation, followed by electrodeposition and molding. The synchrotron radiation allows very narrow features with significant depths to be created due to the short wavelengths and high energies. In these high aspect ratio structures, the sidewalls are very parallel and smooth due to the collimated nature and wavelength of the light. The mask used for LIGA is gold patterned on a relatively x-ray transparent substrate. At Sandia, we typically use a 100 micron thick, three- or four-inch diameter silicon wafer as the substrate material. The resist material is polymethylmethacrylate (PMMA), usually attached to a conductive substrate. During the x-ray exposure, the mask and photoresist are scanned in the x-ray beam until a minimum dose is achieved in the PMMA nearest the substrate. This process can take anywhere from one to twenty hours depending on the thickness of the PMMA, the mask substrate materials, the beam energy, and other considerations. The exposure process causes the PMMA bonds to be severed in the areas irradiated but remain unbroken in the areas under the gold mask. When placed in a chemical developer, the areas with the shorter bond lengths that have been exposed to x-rays are washed away creating a pattern in the PMMA. Once the pattern is created, the PMMA on the conductive substrate can be placed in an electrodeposition bath and metal deposited from the bottom of the mold to the top. The electrodeposition process typically does not proceed in at a uniform rate across the wafer due to the geometric variations that are present. Therefore, after electrodeposition a lapping and polishing step occurs. Once metal deposition is complete and the wafer polished, the PMMA can be dissolved. This metal pattern(s) created can then be used as a tool insert for replication, as metal microstructures on a substrate, or the metal parts can be separated from the substrates and used as freestanding structures.

LIGA: LIthographie, Galvanoformung, Abformung
(Lithography, Electrodeposition, Molding)

- X-rays from a synchrotron are incident on a mask patterned with high Z absorbers. X-rays are used to expose a pattern in PMMA, normally supported on a metalized substrate.

- The PMMA is chemically developed to create a high aspect ratio, parallel wall mold.

- A metal or alloy is electroplated in the PMMA mold to create a metal micropart.

- The PMMA is dissolved leaving a three dimensional metal micropart. These microparts can be separated from the base plate if desired, or the electroplated part can be used as a mold insert.

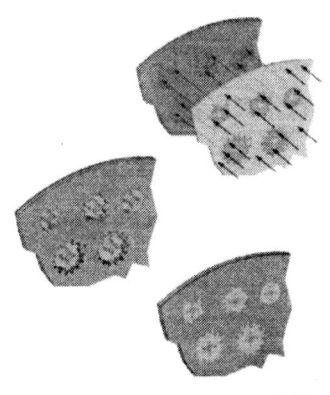

LIGA-like Processes

Since the x-ray lithography step requires synchrotron radiation and therefore access to machines, other lab-based processes have been developed to create LIGA-like fabrication techniques. In these alternative processes, the x-ray lithography step is replaced with another method of creating a non-conducting pattern on a conducting substrate. The subsequent LIGA processing steps can then be followed more-or-less directly. The three most promising pattern- generating techniques used today are thick UV lithography, deep reactive ion etching, and laser machining. In thick UV lithography, a lab-based UV light source is used with a thick UV resist rather than the x-ray source and x-ray resist. While this technique cannot produce feature sizes as small or aspect ratios as large as x-ray LIGA, it is faster and more cost-effective. Thick UV lithography, sometimes called UV-LIGA, is adequate for many geometries and applications. UV-LIGA is limited to thickness of around 500 microns with commercial resist, and aspect ratios about 10-15. Resist modifications have been accomplished at Sandia in order to achieve thickness of 1 mm. SU8 resist is the resist of choice for most UV-LIGA applications.

Deep reactive ion etching is another relatively fast method to obtain molds for subsequent electrodeposition. The exploration of deep reactive ion etching, a non-lithographic method, for non-conductive molds is still in its infancy. However, silicon (semi-conducting) mold etching is quite well developed. Silicon can be used as a patterned material for electroplating if the aspect ratios remain low (less than around 5). If the aspect ratios are too high, the plating that takes place on the semiconductor walls will fill in the pattern prior to complete mold filling. The sidewalls of this technique are significantly rougher than lithographic techniques.

Finally, laser machining is effective at patterning polymers and can be used to create patterns for subsequent electroplating. The limits here are associated with geometric limits and the speed for the laser machining.

- Thick UV resists
    - Relies on i-line lithography
    - Commercially available resists on market (both positive and negative)
    - Generally limited to depths of several hundred microns
    - Sandia has developed capability to use up to 700 microns
- Deep Reactive Ion Etching
    - Relatively quick approach to obtaining high aspect ratio structures
    - (2-6 microns/minute for silicon etching)
    - Geometric variations cause process optimization difficulties
    - Sidewall roughness greater than with lithographic processes
- Laser machining

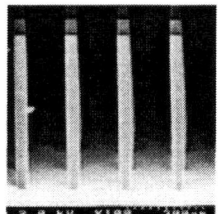

A modified SU-8 resist pattern showing 40 micron lines with a 3:1 pitch, 700 microns tall.

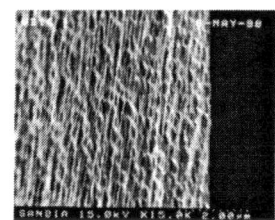

Sidewall of plasma etched feature.

Metal LIGA Prototypes

Many prototype metal parts have been fabricated at Sandia as shown here. These images include several types of gear trains used with electromagnetic actuation to create relatively high torque motors. Also shown is an acceleration sensing device and pieces used to study wear on LIGA parts.

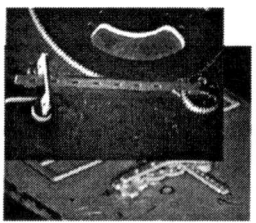

The LIGA technique is most powerful if the metal wafer created is used as a master to create a polymer replicate or mold. Polymer replication is usually accomplished through an injection molding or hot embossing, process. In the injection molding process, the LIGA master is put in the machine and then a high-pressure injection of molten polymer is used to fill the mold. After cooling, the polymer replicate can be removed from the LIGA master. This technique is used in the compact disc industry to stamp polymer discs from nickel masters. It is very fast, with cycle times on the order of a minute. Hot embossing is a process where the LIGA master is heated simultaneously with a polymer sheet to just above the polymer glass transition temperature. The LIGA master is then pressed into the polymer. The polymer and mold and cooled to below the glass transition temperature and separated. This process has excellent reproduction capabilities, however the cycle time is higher and the process is not suitable for very high aspect ratios.

Once a polymer is created, the mold can then used to create additional metal structures using electrodeposition, ceramic or metal microstructures by nanoparticle dry pressing, or polymer microstructures by additional polymer molding techniques. All these techniques have been successfully demonstrated.

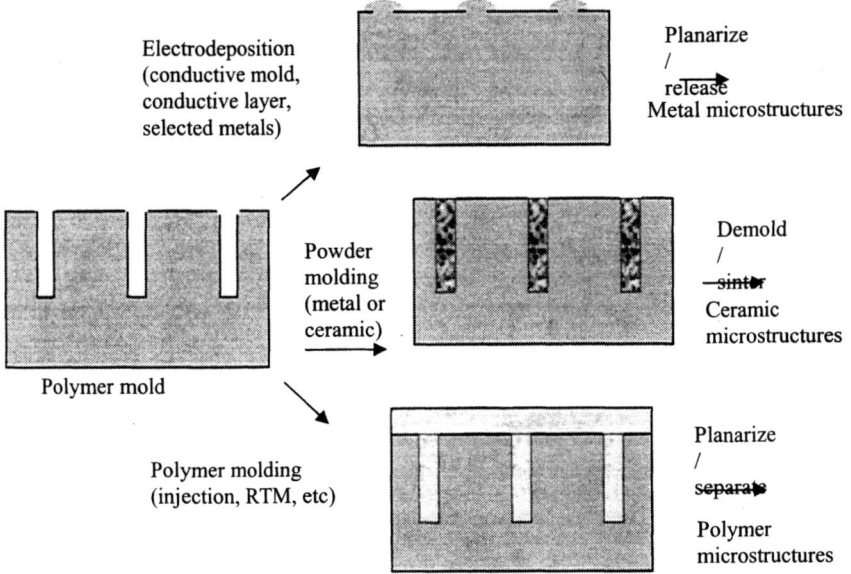

CURRENT RESEARCH TO IMPROVE AND EXPAND LIGA

Current research on LIGA at Sandia includes establishing a wider range of electroplated materials and an understanding of the process/structure/property relationships of the materials. In addition more complete process understanding to improve tolerances is underway. Finally, the use of replication technologies to both increase the range of materials and lower the cost of LIGA is an active area of research.

The materials electrodeposited in LIGA structures today consist primarily of nickel and nickel binary alloys, specifically nickel-iron and nickel-cobalt. Copper and gold are also sometimes used. It has been seen that the current density and detailed electrodeposition chemistries can dramatically affect the microstructure and properties of the electrodeposited materials.

Because of the measured large coupling in the process, structure, and properties, current work involves developing new electrodeposition chemistries optimized for use in small structures, examining the microstructure of the electroplated metal using electron microscopy, and then measuring properties such as strength, plasticity, thermal and electrical conductivity, wear, fatigue, and other relevant parameters. Other characterization necessary include surface roughness and dimensional characterization.

Furthermore, if these LIGA microdevices are to be used in high value applications and expected to last over a long period of time, the effects of aging must also be examined. Sandia has on-going activities to quantify corrosion effects on LIGA electroplated materials as well as to evaluate any changes in grain structure and/or properties that may take place as a function of time.

Develop new electrodeposition chemistries → Examine grain structure

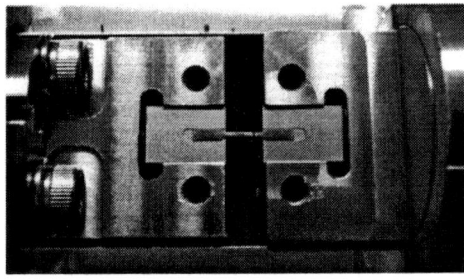

← Age and repeat

Measure properties of plated materials

CONCLUDING REMARKS

LIGA is a well-suited technique to produce microstructures for metals, plastics, and ceramics and is gaining in acceptance due to continued fundamental understanding, successful extension to lab-based pattern generation, and the need for materials other than silicon at the microscale. Today, LIGA free-standing metal parts and LIGA masters can be fabricated. Additionally, replication of the LIGA masters can be successfully accomplished using polymer molding, and nanoparticle pressing.

LIGA offers alternatives for micromachining that are distinct from silicon surface micromachining, bulk silicon micromachining, and state-of-the-art precision machining. Bulk silicon micromachining is a low cost process that produces microstructures made from silicon only, and also has relatively large tolerances. Silicon surface micromachining borrows processing techniques from the integrated circuit industry therefore producing exacting tolerances and dimensions. Limits of silicon surface micromachining include the materials set (silicon and limited coatings), as well as thickness (usually less than 20 microns). One significant advantage of silicon surface micromachining compared to the other techniques is the ability to deploy sacrificial layers so that microsystems can be pre-assembled during processing. At the opposite extreme in microscale fabrication, precision machining techniques such as wire electrodischarge machining is capable of producing fully three-dimensional parts from a large variety of materials. The limit in electrodischarge machining and other "cutting" techniques include smallest feature definition as well as sidewall roughness. The uniqueness LIGA offers includes the material set as well as the sidewall smoothness, sidewall straightness, and aspect ratio.

ACKNOWLEDGEMENTS:

Sandia is a multiprogram laboratory operated by Sandia Corporation for the United States Department of Energy under contract DE-AC04-94AL85000.

REFERENCES

For more information on LIGA visit the following websites: www.sandialiga.org; www.fzk.de; www.camd.lsu.edu; www.dl.ac.uk/LIGA; www.imm.mediadialog24.de; http://syli34.physik.uni-gonn.de/rtlplus.html; http://microfab.lure.u-psud.fr.

"Windowtron" RF Breakdown Studies at SLAC

L. Laurent*,†, G. Caryotakis†, F. Glendinning†
N.C. Luhmann, Jr*., C. Pearson†, G. Scheitrum†, D. Sprehn†

*University of California Davis, Davis, CA 95616
†Stanford Linear Accelerator Center, Menlo Park, CA 94025

Abstract. This paper compares the experimental breakdown results for copper cavities having different external Q-factors (Q_{ext}). The cavity with the lower Q_{ext} has more available energy to fuel a discharge. Factors such as location, number, and distribution of breakdown sites will be presented. The preliminary experimental results of high peak power processing at a reduced pulse length are also presented as a potential means of achieving high field gradients with reduced surface damage. In an effort to increase the breakdown threshold, alternative materials to copper are under investigation. GlidCop® AL-25 and Stainless Steel 317L have recently been tested and are compared to previous copper cavity experiments.

INTRODUCTION

The need to understand the fundamental mechanisms of rf breakdown has been driven by next generation accelerators and sources that will operate at higher frequencies, requiring smaller structures and higher field gradients. Finding methods to achieve these higher gradients and to minimize rf processing time have been the focal point of this study.

In 1997, an X-band TM_{020} cavity with demountable high gradient surfaces was developed to investigate rf breakdown. Initial research topics focused on microparticle contamination, vacuum conditions, grain boundaries, pulse length, and coatings. The reader is referred to Ref. [1,2] for a more detailed account of these experiments.

During the past decade, prototype NLC X-band traveling wave accelerator structures have been tested with breakdown damage found to occur at much lower gradients than expected. A group velocity (v_g) dependence on breakdown was proposed because the rf power required to achieve a given gradient increases with v_g [3]. The group velocity varies linearly with coupling (Q_o/Q_{ext}). To understand the relationship between coupling and breakdown, two cavities having different external Q values have been tested.

The preliminary experimental results of high peak power processing at a reduced pulse length are also presented as a potential means of achieving high field gradients with reduced surface damage. In earlier work [1], it was noted that during rf processing at a single pulse length, impurities on the surface either melted, evaporated, exploded, or remained unaffected. Limiting the energy by initially processing at a reduced pulse duration seeks to remove field enhancements without causing surface damage.

Previously, all the breakdown cavity structures have been fabricated from OFE copper. In an effort to increase the breakdown threshold, different materials are being explored as an alternative to copper. GlidCop® AL-25 and Stainless Steel 317L have recently been tested. Difficulties in achieving a smooth, particle-free surface have exemplified the need to conduct a preliminary investigation, involving machining and etching techniques, when working with new materials.

EXPERIMENTAL SETUP

The rf breakdown experiments are conducted inside a field enhanced transmission cavity operating in the TM_{020} mode (Fig. 1). Two cavities with different external Q's have been manufactured for this research. The field strength is proportional to the square root of the power radiated out of the cavity and the external Q (Q_{ext}) of the output iris. The high Q_{ext} cavity has a Q_{ext} of 2085 and the low Q_{ext} cavity has a Q_{ext} of 190. Both cavities have two demountable nose tips designed to facilitate nondestructive analysis of the high gradient surface area. The first radial field maximum for the TM_{020} mode is located at the center, which has the appearance of a button in the photograph (diameter ~ 0.5 cm). The gap distance is 1.9mm. The nose tips can be made from various materials, and different cleaning and machining techniques can be applied. After each experiment, the cavity noses are removed and analyzed by a scanning electron microscope with energy dispersive x-ray analysis (SEM/EDX).

The cavity is situated between two TE_{11} ceramic windows, and all components between the two windows comprise the "windowtron" (Fig. 2). The windowtron

FIGURE 1. Photograph of the TM_{020} cavity with one of the removable cavity noses shown in the foreground.

vacuum is isolated from the rf source enabling the cavity and components within the windowtron to be baked out as a complete assembly prior to rf testing. After vacuum bake, initial pressures of 10^{-11} Torr have been achieved. The windowtron input source is a 50 MW X-band klystron operating at 11.424 GHz with a pulse repetition rate of 60 Hz. The klystron output power is coupled into a 3-dB magic-tee to protect the klystron from reflected power during a breakdown event in the cavity. RF power is measured at the input and output of the cavity by two 55-dB crossguide directional couplers.

In 1998, as part of this breakdown study [1], it was found that vacuum firing the cavity noses prior to rf testing significantly reduced the number of breakdowns. Furthermore, it was shown that for nonvacuum fired cavity noses, trapped gases at grain boundaries were liberated during rf processing and was the initial cause of breakdown. For this reason, all cavity nose pairs are vacuum fired prior to installing them into the cavity. Once the noses are in the cavity, the windowtron is either heat tape baked (150°C) or vacuum baked (450°C) depending on the experiment.

In past experiments, detecting breakdown was based on a sudden increase in gas pressure or visible light. A more precise method to detect breakdown is to analyze the missing energy in each rf pulse. In all the following experiments, each rf pulse is analyzed by data acquisition software, and if more than 10% of the energy is missing, a software generated signal disables the rf preventing the next pulse from propagating.

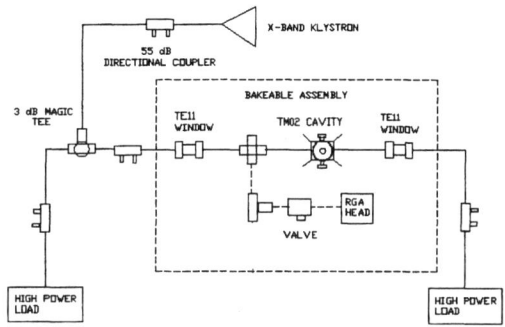

FIGURE 2. Experimental setup for rf breakdown testing inside a windowtron.

HIGH Q AND LOW Q CAVITY BREAKDOWN COMPARISON

To identify if a relationship exists between coupling and vacuum breakdown, four low Q_{ext} cavity and two high Q_{ext} cavity experiments have been conducted while maintaining a strict rf processing protocol. The cavities were rf processed at a pulse length of 240ns. The power was applied with progressively higher rf field levels in small discrete steps up to 200 MV/m. A ten-minute processing time was allowed at each step. Once breakdown is detected, the data acquisition system disables the rf drive. Prior to resuming operation, the power is lowered and gradually increased to a field gradient below the breakdown point for additional processing.

Comparison between the low Q_{ext} (LQ) and high Q_{ext} (HQ) experiments revealed that substantially more surface damage occurred on the cavity noses processed in the low Q_{ext} cavity. This can be seen from the first four white columns shown in Fig. 3a. The white columns represent the total number of breakdown sites visually detected by

SEM for each cavity nose pair. The last two white columns are from the two high Q_{ext} cavity experiments, showing substantially less damage. Additional experiments with the high Q_{ext} cavity are scheduled to provide a more reliable statistical comparison. The cavity noses for each of these experiments were fabricated from OFE copper. All the windowtron structures were initially vacuum baked (450°C) except for LQ4 and HQ2, which were heat tape baked (150°C). The difficulty in understanding the fundamental mechanisms involved in vacuum breakdown is reflected in the large variance in surface damage when comparing the first three low Q_{ext} cavity experiments. These experiments were fabricated from the same lot of material, vacuum baked and rf processed with the same processing protocol. In the low Q_{ext} cavity experiments, the windowtron that was heat tape baked (LQ4), had more surface damage; however, further experiments would be necessary to draw conclusions. In fact, incongruently, of the two high Q_{ext} cavity experiments, the one that was heat tape baked (HQ2), had less surface damage. Additional experiments would be necessary to determine if this was an anomaly or if there is a correlation between bakeout temperature and breakdown.

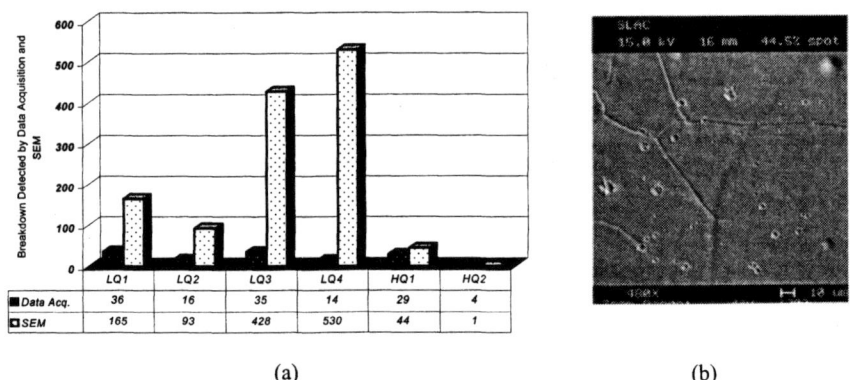

(a)   (b)

FIGURE 3. (a) More surface damage (white columns) was visibly observed on the cavity noses processed in the four low Q_{ext} (LQ) cavities compared to the two high Q_{ext} (HQ) cavities. The number of breakdown events detected while processing (black columns) were significantly lower than the number of craters detected. (b) A SEM photograph showing how breakdown sites appear in clusters in the low Q_{ext} cavity experiments.

The black columns in Fig. 3a represent the total number of pulses with more than 10% missing energy detected while rf processing. There is not a one-to-one correspondence between the number of pulses detected with missing energy and the visible number of breakdown craters observed on the surface. This may be due to more than one breakdown event occurring during a single rf pulse. Another possible cause may be due to breakdown events having less than 10% missing energy going undetected by our data acquisition system. The average field gradient for the first breakdown event detected while processing was lower for the four low Q_{ext}

experiments (111 MV/m) than the average for the two high Q_{ext} experiments (137MV/m).

Although the number of breakdowns craters varied significantly in the low Q_{ext} cavity experiments, a further investigation showed that the breakdown sites tended to be in groups (Fig. 3b), while the breakdown sites observed on the high Q_{ext} cavity noses tended to be single, isolated events. It is suspected that once a breakdown is initiated, the additional energy available in the low Q_{ext} cavity melts more copper, displacing droplets of molten copper to the surrounding area. These droplets can provide field enhancement sites leading to further breakdown.

After reaching the maximum gradient (200 MV/m), each of the cavity noses was rf processed for an additional 40 hours at 150MV/m. In all cases, there was no additional breakdown detected at the reduced gradient. This step showed that once a cavity had been processed above a given gradient, a reduction in field gradient by 25% provides a safe operating gradient.

There was no correlation in the location or the number of breakdown sites between the cavity nose pairs. The number of breakdown craters detected on each nose for each of the six experiments is shown in Fig. 4a. The x-axis depicts the cavity experiment number, and the height of the columns (y-axis) gives the number of breakdown sites detected by SEM for each cavity nose pair (z-axis). The asymmetry of breakdown sites is particularly evident in experiment LQ2 where only one breakdown site was detected on cavity nose A, while 92 breakdown sites were detected on the opposing cavity nose B.

Breakdown images integrated over a single rf pulse were captured with a CCD camera (Fig. 4b). The nose pairs are vertically aligned (Fig. 4b-insert). The reflection observed on the insert is due to the external light source used to image the noses prior to breakdown. During breakdown, a plasma formation is detected localized near the surface with no visible evidence of an arc traversing the gap. If breakdown is a single surface phenomenon, this would also explain breakdown asymmetry between the cavity noses.

High Peak Power Narrow Pulse Processing

High peak power processing at a reduced pulse duration is an rf conditioning technique that limits the available energy to be dissipated during breakdown. The reduction in field emission observed during rf processing suggests that emission

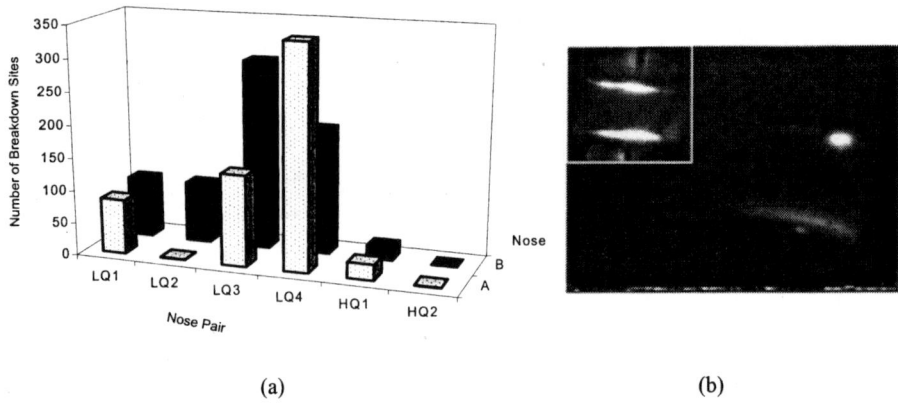

FIGURE 4. (a) The number and location of breakdown sites between adjacent noses (A, B) in each experiment showed no correlation. (b) Plasma formation captured by a CCD camera during breakdown. Insert shows orientation of cavity noses.

sites can be removed and/or their geometry modified. By initially processing at a shorter pulse duration, the intent is to evaporate surface irregularities or potential emitter sites at a lower energy that may otherwise explode producing additional unstable field emitters.

Two experiments using high peak power narrow pulse processing have been conducted. One experiment was conducted in the low Q_{ext} cavity and the other utilized the high Q_{ext} cavity. Both windowtron structures were vacuum baked at 450°C. The starting pulse duration was 80ns, and was incrementally increased up to 240ns. The same basic rf processing protocol used in the previous six experiments was followed at each pulse length. The only exception was that the peak field gradient was extended to 300 MV/m at the shortest pulse length and was incrementally reduced to 200 MV/m at the longest pulse duration (240ns). The final field gradient and pulse length were selected to be consistent with the previous six experiments. The pulse lengths and corresponding peak field gradients are shown in Fig. 5a.

In the high Q_{ext} cavity, there were 34 breakdowns detected by the data acquisition system while processing with a 80ns pulse length (Fig. 5b). After increasing the pulse duration to 130ns, two more breakdowns were detected. No additional breakdowns were detected at the two longer pulse lengths (180ns, 240ns). In the low Q_{ext} cavity, there were 77 breakdowns detected which all occurred at the shortest pulse length (80ns). The low Q_{ext} cavity had substantially more breakdown on the surface due to some surface damage (dents) acquired prior to rf testing. Numerous breakdown sites were observed along sharp edges of these dents. In spite of the pre-damaged surface, it was interesting that the results for these two experiments were consistent. They both showed that processing at shorter pulse lengths does condition a surface resulting in few breakdowns at longer pulse lengths. The difficulty in this processing technique is in determining the optimum peak field gradient for each corresponding pulse length. Most of the breakdown occurred at the shortest pulse duration while processing between 250 MV/m and 300 MV/m. It is believed that if the peak field strength for each pulse length could be optimized, less surface damage would be likely due to the reduction in available energy to fuel a discharge.

Pulse Length (ns)	Peak Field Gradient (MV/m)
80	300
130	275
180	250
240	200

(a)

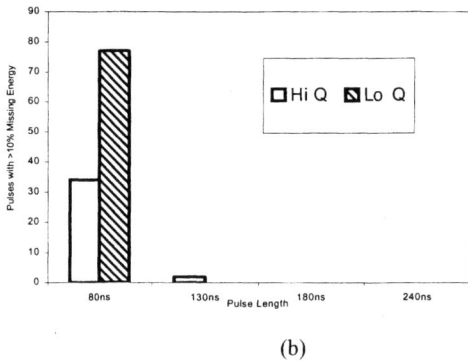

(b)

FIGURE 5. (a) Pulse lengths and maximum field gradients selected for high peak power narrow pulse processing experiments. (b) Both the high Q_{ext} and low Q_{ext} cavity experiments showed that processing with a shorter pulse length can reduce the number of breakdowns at longer pulse lengths.

Sustaining a Plasma with Low Power Rf

Recent experiments in accelerator structures at the Stanford Linear Accelerator Center have shown that after breakdown, light measured with a photomultiplier tube (PMT) extends long after the input rf pulse. It was postulated that this was due to residual stored energy. An attempt was made with the windowtron facility to experimentally quantify the energy necessary to sustain a plasma once breakdown has occurred. To accomplish this, the low Q_{ext} cavity was driven with an 80ns rf pulse riding on a variable low power shoulder. The power level for the low power shoulder was varied with as little as 28% of the peak power being sufficient to sustain the plasma. A further reduction in the power level was not available without completely eliminating the low power shoulder. The light monitored by the PMT closely followed the duration of the low power shoulder (~2µs) once breakdown was initiated by the peak rf field (Fig. 6).

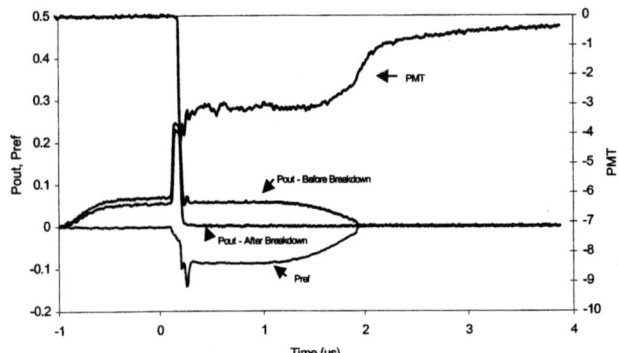

FIGURE 6. The power at the output of the cavity (P_{out}) is shown before and after breakdown. After breakdown was initiated by the peak power (P_{pk}), a low power shoulder (28% P_{pk}) was sufficient to sustain the plasma (detected by a PMT) for the duration of the shoulder (~2µs).

MATERIALS STUDY

Copper is widely employed in high power microwave devices and components because of its high electrical and thermal conductivities. However, one drawback of pure copper is its low tensile and yield strengths at room temperature and at elevated temperatures. The current rf breakdown study has extended the testing of copper cavity noses to include GlidCop® (AL-25) and Stainless Steel (317L), which are harder materials and have higher yield strengths compared to copper. Our research has found that studying different materials is a complicated process. A standard cleaning, etching, and machining process cannot be translated from one material to the next. When working with new materials, a significant investment of time and funding is required to study techniques that can provide a smooth, particle-free surface finish.

GlidCop®

GlidCop® consists of a pure copper matrix containing finely dispersed sub-microscopic alumina particles. The aluminum oxide particles act as a barrier to dislocation and grain boundary movement thereby retarding recrystallization, which leads to softening. This study investigated GlidCop® AL-25, which is a low alumina content grade that is resistant to softening up to temperatures close to the melting point of copper. GlidCop® is primarily composed of pure copper so the conductivity properties closely match those of pure copper. A low oxygen content grade was utilized to prevent the development of porosity during vacuum brazing.

Two experiments were conducted with GlidCop® cavity nose tips. Prior to rf testing, an SEM photograph was taken after chemically etching the cavity noses (Fig. 7a). Numerous submicron size aluminum oxide particles were observed on the surface due to the preferential etching of copper. The circle in the photograph is provided as a

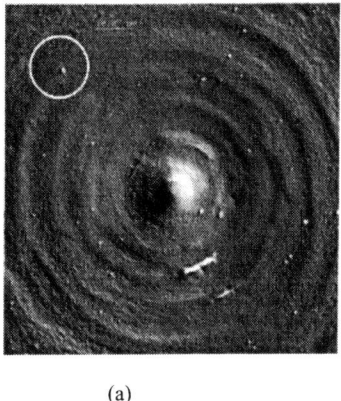 

(a)   (b)

FIGURE 7. SEM photograph showing GlidCop® cavity noses (a) before and (b) after rf processing up to 200 MV/m (Mag. 1000x). Circle encloses one of the many submicron alumina particles found on the surface after etching. Note change in surface texture after rf processing.

pointer to one of the particles. There were too many particles on the surface to conveniently map all of them prior to rf testing; however, approximately 50 particles were located and documented for subsequent reference. In addition to the alumina particles, the surface finish was not as smooth or uniform relative to the copper cavity noses.

The GlidCop® noses were installed in the low Q_{ext} cavity and heat tape baked. The same processing protocol was employed as on the six copper cavity nose experiments previously discussed. The first breakdown detected by the data acquisition system occurred at 96 MV/m. While processing up to 200 MV/m, there were 108 breakdowns detected. This number was significantly larger than observed in the copper cavity nose experiments. After rf testing, the noses were SEM analyzed (Fig. 7b). Several breakdown sites appear on and around the location of the elongated protrusion visible in Fig. 7a, located below the center of the cavity nose. Chemical analysis revealed this section to be copper. The aluminum oxide particles that were previously mapped remained on the surface, with the exception of those near the protrusion. The surface texture changed significantly after rf processing giving it a matted appearance. The cause of this is unclear, and has never been observed in the experiments using pure copper.

There were too many overlapping breakdown sites on the GlidCop® noses, thereby preventing a numeric comparison to the copper noses; however, the grouping of breakdown sites is clearly visible in Fig. 8 (Magnification: 44x). A second set of GlidCop® noses were also tested with comparable results.

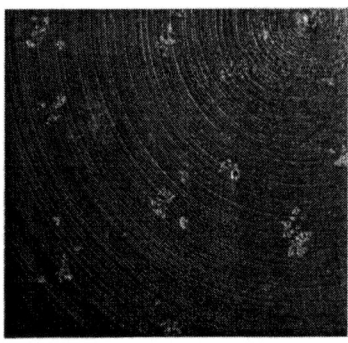

FIGURE 8. Breakdown site groupings were prominent in this low Q_{ext} cavity experiment. (Magnification: 44x)

Stainless Steel 317L

One experiment was conducted with stainless steel (SST) 317L cavity nose tips. SST 317L is a nonmagnetic alloy with a low carbon (0.03% max) content. To avoid damaging copper sections on the cavity noses, a copper etching solution with a longer etch time was initially employed; however, sharp edges remained on the surface (Fig. 9). The cavity noses were rf tested with breakdowns occurring at field strengths 5-6 times lower (20 MV/m) than on the copper cavity noses. High field concentrations can occur at relatively low voltages when surface irregularities exist. This exemplified the importance of smooth surfaces as a prerequisite in achieving high gradients.

The cavity noses were removed and electropolished. Electropolishing is a surface finishing technique for stainless steel (as well as many other alloys) that is capable of producing a very smooth surface. During electropolishing, the cavity noses formed the anode and were placed in an electrolytic bath of acidic chemical solution. Large DC currents flow from the anodic cavity nose and a metal cathode. Electric fields focus on the microscopic peaks, enhancing the local material removal rate over that of valleys.

An SEM photograph taken at the center of the cavity noses is shown before (Fig. 10a) and after (Fig. 10b) electropolishing. A closer examination of the surface after electropolishing revealed approximately 45 alumina particles residing on the surface of one cavity nose. All of the particles were approximately 10μm in size and each particle was found to be centered inside a depression (Fig. 11a). A possible explanation for the source of alumina particles is the use of aluminum as a deoxidizer in the manufacturing process of stainless steel [4].

The stainless steel noses were inserted into the high Q_{ext} cavity and heat tape baked. The same processing protocol was followed as on the six copper cavity nose experiments.

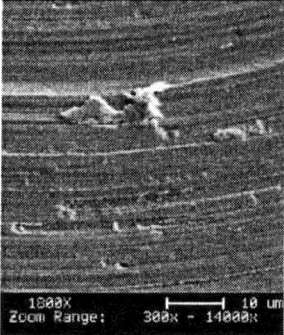

FIGURE 9. Sharp edges on the stainless steel cavity noses were detected after chemical etching. Breakdown occurred at very low gradients (20 MV/m).

(a) (b)

FIGURE 10. Stainless steel cavity nose before (a) and after (b) electropolishing. The photographs are taken at the center of the cavity nose tip at a magnification of 1000x.

The first breakdown occurred at 100 MV/m, and there were a total of 43 breakdowns detected while rf processing.

SEM analysis of the premapped cavity nose revealed ten groupings of breakdown sites. Seven of these groups corresponded to previously mapped alumina particles. The alumina particle in Fig. 11b is the same particle shown in Fig. 11a after processing up to 200 MV/m. Groupings of breakdown sites were located near particle locations (Fig. 11b,c). Alumina was detected only inside the crater directly beneath the particle, while stainless steel was all that was detected at the breakdown sites located near the particles. The grouping of breakdown sites is unusual in the high Q_{ext} cavity. This may suggest that impurities may have been a factor in the three breakdown locations that could not be correlated to premapped alumina particles. Unlike the Cornell and CERN studies [5,6] on niobium cavities, our research [1] indicates that on copper surfaces, particles smaller than 5μm do not play a significant role in rf breakdown. However, in this experiment the particles were larger (10μm) and did contribute in the breakdown process.

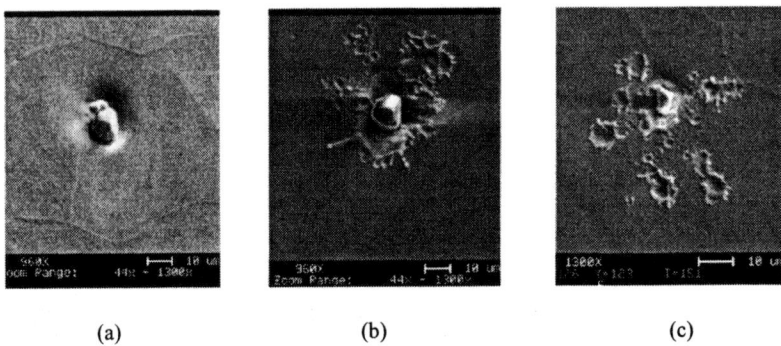

(a) (b) (c)

FIGURE 11. Numerous alumina particles were detected after electropolishing. An SEM photograph showing one alumina particle (a) before and (b) after rf processing. (c) Groupings of breakdown sites were found located near seven of the premapped alumina particle locations.

CONCLUSIONS

A breakdown comparison between a high Q and low Q cavity revealed that surface damage was more severe in the low Q structure and that breakdown sites tend to be in groupings. In contrast, the breakdown sites in the high Q cavity typically were single, isolated events. These results were consistent with prototype NLC accelerator structures, where the maximum field gradient appears to vary almost linearly with the inverse of the group velocity. It was also shown that low levels of residual energy are capable of sustaining a plasma once breakdown occurs long after the rf pulse.

A preliminary investigation of high peak power processing at a reduced pulse duration revealed that most of the breakdown occurs at the shortest pulse length. It is believed that if this process were optimized, less surface damage is likely due to the reduction in available energy to fuel a discharge.

In an effort to increase the breakdown threshold, different materials are being explored as an alternative to copper. This paper has discussed the difficulties which may be encountered on a microscopic level when working with different materials.

ACKNOWLEDGEMENTS

The authors wish to express appreciation and acknowledge the efforts of the employees in the klystron department at SLAC for their contributions to this research. This work has been funded by the AFOSR under Grant F49620-95-1-0253 (MURI), and by the Department of Energy Contract #DE-AC02-76SF00515.

REFERENCES

1. Laurent L., Scheitrum, G., Caryotakis, G., Vlieks, A., Luhmann Jr., N.C., *RF Breakdown Experiments at SLAC*, RF98 Workshop, Monterey, CA, October 5-9, 1998.
2. Laurent, L., Caryotakis, G., Scheitrum, G., Sprehn, D., Luhmann Jr., N.C., *Pulsed RF Breakdown Studies*, SPIE Intense Microwave Pulses VII, Orlando, Florida, April 24-26, 2000.
3. Adolphsen, C., Baumgartner, W., et. al., *Processing Studies of X-Band Accelerator Structures at the NLCTA*, Particle Accel. Conf., ROAA003, Chicago, June 18-22, 2001.
4. Francis, J., SLAC, *Private Communication*, 2001.
5. Knobloch, J., *Advanced Thermometry Studies of Superconducting RF Cavities*, PhD Thesis, Cornell University, 1997.
6. Niedermann, P., *Experiments on Enhanced Field Emission*, PhD Thesis, University of Geneva, 1986.

RF Breakdown in High Vacuum Multimegawatt X-band Structures

Valery A. Dolgashev and Sami G. Tantawi

SLAC, Stanford, CA 94309, USA

Abstract. Increasing the power handling capabilities of rf components is an important issue for the design of rf accelerators and rf sources. RF breakdown is a phenomena that limit the high power performance. A major concern is the damage that can occur in rf components from breakdown. To better understand this damage, we have studied rf breakdown in a rectangular waveguide experimentally and theoretically. The breakdown process in a waveguide is both easier to measure and simulate than breakdown in a complex geometry such as an accelerating structure. We used a particle tracking code and a Particle-In-Cell code to model the breakdown behavior. Models developed for the waveguide were applied to the breakdown in accelerating structures. RF breakdown in traveling wave and standing wave accelerating structures was simulated. We compare the experimental data with results of the simulations for the accelerating structures.

INTRODUCTION

Accelerating gradient is one of the crucial parameters affecting design, construction and cost of next-generation linear accelerators. For a specified final energy, the gradient sets the accelerator length, and for a given accelerating structure and pulse repetition rate it determines power consumption. Accelerating gradients on the order of 100 MV/m have been reached in short ($\sim$ 20 cm) standing wave and traveling wave X-band accelerating structures [1, 2, 3]. But recent experiments have shown damage to traveling wave accelerating structures at gradients as low as 50 MV/m after 1000 hours of operation [4]. RF breakdown is a probable cause of this damage. An extensive experimental and theoretical program to determine a safe operating gradient for the Next Linear Collider (NLC) is under way in SLAC. The present work is a part of that program.

RF breakdown

We define rf breakdown as a phenomenon that abruptly and significantly changes transmission and reflection of the rf power directed to the structure under test. We distinguish breakdown from field emission and dark current. Dark currents have reproducible and monotonic (with respect to input power) behavior in spite their random space-time origin. There is evidence [4] that rf breakdown can damage the structure. RF breakdown is a complex phenomenon and its physics is yet to be understood. It includes the rf driven interaction of electrons, ions and neutral atoms, heating and melting of the metal surface *etc*. The short time scale of the breakdown ($\sim$ 5 ns–10 μs), its unpredictable starting time, and the random location of the breakdown site make it very difficult to observe its microscopic behavior. In contrast, we routinely detect and record external macroscopic parameters such as incident rf power, power reflected from the breakdown site, and power transmitted through it. We use other parameters, such as emitted light, X-rays and harmonics of the working frequency to obtain more information about the physics of the breakdown phenomenon. Other parameters are the electron currents that exit from the beam pipe of the accelerating structure.

Simulations

We use a simplified physical model to simulate the breakdown. We assume that part of the structure surface starts emitting electron and ion currents at a predetermined time (the physics of this emission are not considered here). We adjust the emission models to resemble observed parameters (transmitted and reflected rf) of the experimental data. We have chosen a commercial Particle-In-Cell (PIC) code MAGIC [6] for these simulations. This code is used at SLAC for 2D and 3D simulation of klystrons. Such features of the code as the capability for input of realistic geometries and developed diagnostics (of particle and field parameters) make it very useful for the simulations. Unfortunately, the chosen code does not let us use small emission areas. Traces left by breakdowns on metal surfaces have a size $\sim 10 - 100$ μm and the smallest mesh size in the PIC code is ~ 1 mm. Another limitation is the maximum size of the simulated structure. We have simulated up to 8 cells (~ 10 cm) of 11.4 GHz standing wave accelerating structure in 2D and 5 cells (~ 5 cm) of traveling wave structure in 3D, while real structures have lengths up to 1.8 m. The simulations were guided by comparison with measurements of breakdown in a simple rectangular waveguide. The breakdown process in a waveguide is both easier to measure and to simulate than breakdown in the complex geometry of an accelerating structure.

BREAKDOWN IN A WAVEGUIDE

Experiment

As part of the breakdown research program we studied rf breakdown in a rectangular waveguide.

Its width was reduced to 1.33 cm (in comparison with a width of 2.29 cm for WR90) over a length of 6 cm in order to enhance the electric field by lowering the group velocity to 0.18c, and force the breakdown to occur in a this area. The height of the waveguide is 1.02 cm. We subjected the waveguide to rf power up to 120 MW with pulse widths up to 1.2 μs. We recorded incident, transmitted and reflected rf power; intensity of light emission and its spectrum; intensity of X-rays; and harmonics of the working frequency of 11.424 GHz. Harmonics were present in the klystron output as well as being generated by rf breakdown. We list some general conclusions about breakdown behavior deduced from the results of this experiment:

1. Transmitted power has a repeatable shape: it drops off to zero with an amplitude proportional to $e^{\frac{-(t-t_s)^2}{2\tau^2}}$ (Gaussian-like). Here t_s is breakdown start time. The rf pulse starts at $t = 0$. The range of the drop off time constant τ is between 10 and 200 ns. For a preprocessed waveguide τ is ~ 30 ns. Here preprocessing means

steady running at ~ 100 MW of rf power through the waveguide with a short pulse length ~ 300 ns.
2. Transmission does not recover for several microseconds after the breakdown.
3. During the Gaussian drop off the transmitted and often the reflected signals have a few oscillations.
4. Up to 90% of the incident rf energy is absorbed after $t_s + 2\tau$. Breakdowns in a preprocessed waveguide absorb (on average) less energy than otherwise.
5. RF transmission fully recovers after the main rf pulse has been off for several milliseconds.
6. Light (emitted from the breakdown site) lasts for several microseconds after the rf pulse.
7. Spectral lines of the light are mostly from neutral copper atoms (Cu I) with traces of Cu II ions and hydrogen.
8. Breakdowns tend to occur on sequences of rf pulses. Subsequent breakdowns most probably have a shorter starting time t_s than the first breakdown of a sequence.
9. Breakdowns at rf high power (~ 100MW) and short pulse length (~ 300 ns) decrease dark currents, and lower power and a longer pulse (more than 400 ns) increases dark currents. We think that the level of dark currents indicates the degree of metal surface damage.
10. Breakdown has no detectable effect on laser light (632.8 nm) passing through the waveguide.
11. In most events, the 3rd harmonic (34.272 GHz) signal from the klystron transmitted through the breakdown site is shut off by the breakdown.
12. Breakdown produces a 3rd harmonic of the klystron signal, and, probably, higher harmonics.
13. Characteristic size of the damaged spots on the surface of the waveguide is $\sim$ 10–100 μm.

Simulations

We performed 3D PIC simulations of the breakdown in the waveguide. The waveguide model has the same cross section as the experiment and a length of 6 cm. There is no reflection from the waveguide's ends for the TE01 mode. We simulated breakdown by creating emission spots on the broad walls of the waveguide. We applied space-charge limited emission of electrons. For that we used the built-in code feature EMISSION EXPLOSIVE. For copper ions we used a beam generated at the same area with a predetermined current density and initial velocity distribution (BEAM model in MAGIC). We changed such parameters as size and position of the emitting spots, input power, initial characteristics of the ion beam, and density of a neutral gas. We varied the size of the emitting spots from 1.6 mm $\times$ 1.6 mm up to 1 cm $\times$ 4 cm. From this numerical experiment we came to the conclusions listed below.

1. The major energy exchange between incident rf fields and particles comes from the interaction of the rf electric fields with electrons (not with ions). Electrons cross the

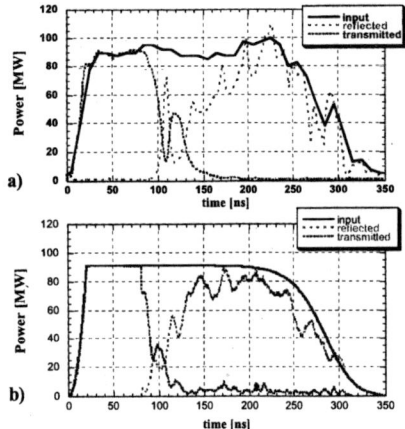

FIGURE 1. Incident, reflected and transmitted rf power for a breakdown in the waveguide. a) a measurement of a typical breakdown in the preprocessed waveguide. b) 3D PIC simulations, with an emitting spot size 4 mm×4 mm, an electron current of 7 kA, and a copper ion current of 30 A.

waveguide in a short time ($\sim$ rf period).

2. The electron current must be several kA to significantly effect the rf power transmission. If we take into account the size of the damaged area, the current density must be in the order of 10^7 A/cm^2. Space charge-limited emission of electrons without ions cannot produce and sustain such current densities.
3. Ion currents must be 10 to 100 A to disrupt transmitted power. An initial energy up to 50 eV does not change the ion dynamics. The space charge fields of the ions compensates the electron space charge fields. This compensation allows the generation of kA of electron current. The time constant of the drop off of the transmitted power τ is 10 to 20 ns and is related mostly to the process of filling the waveguide gap with copper ions.
4. Without electrons the ions do not move significantly during the rf pulse. In the presence of space-charge limited electron flow, the ion beam crosses the waveguide in about 30 ns at 80 MW of input power. The oscillating space charge field of the electrons adds a dc component to the rf electric field that accelerates the ions.
5. A significant portion (50–80%) of the emitted electrons and ions returns to the emitting spot and the surrounding area.
6. During the rf pulse, most electrons and ions are confined to a beam with a cross-section area of about 1 cm^2.
7. The transmitted and reflected power oscillates with a period 10–40 ns, determined by the ion-electron density.
8. The ion-electron current generates harmonics of the working frequency. The perturbation of the incident electric field due to these harmonics is on the order of 10%.
9. Up to 50% of the input power can be absorbed by the ion-electron beam.

10. Up to 75% of the input power was absorbed by the ion-electron beam after we added some effects associated with the interaction of electrons with neutral copper atoms.

A comparison between a signal from an typical actual breakdown and simulation with similar parameters is shown on Fig.1. The experience that we gained using the PIC code to understand waveguide breakdown gives us confidence in applying the same method to study breakdown in accelerating structures.

ACCELERATING STRUCTURES

The characteristics of rf breakdown in traveling wave (TW) accelerating structures [5] are similar to those of waveguide breakdowns. The main similarities are: a drop off to zero of the transmitted rf power in tens of nanoseconds, with up to 80% of the incident rf energy absorbed after breakdown starts. We think that this analogy comes from some common characteristic of the waveguide and TW structures. Both have a broadband frequency response and are designed to transmit rf power. A breakdown that is localized in a limited volume or single cell does not change the ability of the structure to channel power to the breakdown site. We simulated 2D and 3D models of a TW accelerating structure. We used dimensions of a structure with an initial group velocity 0.05c that is currently under high power test at SLAC. We placed the emission spots on the iris of a structure cell. Results of PIC simulations of the accelerating structure lead to conclusions that are very similar to those for waveguides. We add some conclusions related to specifics of the accelerating structure below:

1. There is an asymmetry in the portion of electron current that exits through the beam apertures. Current is more likely to go toward the input coupler from the cell with the emission spot.
2. There is no significant difference between 2D and 3D models. Assuming the same total emission current, reflected power and transmitted rf power behave similarly in both models.
3. Electron current from the spot spreads over the inside surface of the cell, but the major part of the emitted current goes to a small area on an iris opposite to the emitting spot and to a current returning back to the emitting spot.
4. Secondary or back-scattered electrons do not change significantly the behavior of the rf fields.

Breakdown behavior of standing wave (SW) structures [1, 2] is very different from TW structures. In the TW case, a major part of the rf energy is absorbed by breakdown currents; in the SW case rf energy is reflected from the structure. After breakdown starts, reflected energy increases during $\sim$100 ns in TW case, and in $\sim$10 ns in SW case. We simulated a 2D SW structure to find the source of these differences. We used dimensions of the π phase advance SW structure that is currently under high power test at SLAC [5]. The simulated reflected power and a signal from a field probe are shown in Fig. 2. The main results of these simulations are listed below:

1. Electromagnetic fields in the structure collapse just after emission starts. Currents pass across the whole cavity and absorb a major part of the stored rf energy in a few nanoseconds, compared to a filling time of $\sim$100 ns.

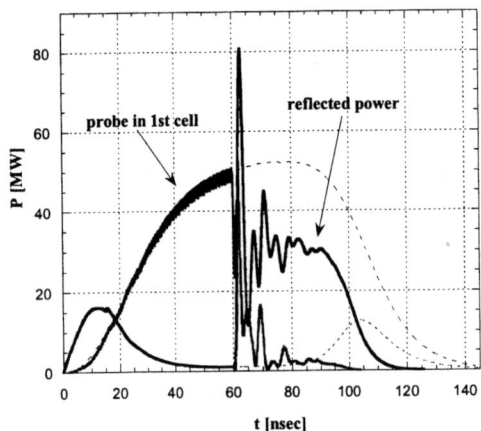

FIGURE 2. Reflected power and power from a simulated probe in the structure *vs.* time. Power from the probe is arbitrarily normalized. Emission starts at 60 ns. Dashed curves — no emission, solid curves — with emission.

2. The currents detune the whole structure, causing the π-resonance to shift from the working frequency. This shift causes rf energy to reflect from input iris of the structure.
3. The increase of the electric field in the cells (due to emitted currents) is generally smaller than for the TW structure.

The main difference between the TW and SW case is in the coupling of the structure cavity to the input waveguide. The goal in TW coupler design is to have a small reflection from the coupler over a wide frequency range ($\frac{\Delta f}{f} \sim 10^{-3}$). In the SW case, the goal is a small reflection from the beam-loaded structure in a narrow frequency range ($\frac{\Delta f}{f} \sim 10^{-4}$). In the TW case breakdown currents on the order of 10 A have a negligible effect on transmission and reflection of rf power, but in the SW case the same current shifts the resonant frequency enough to cause reflection of a major part of the incident rf power. We think that this high sensitivity of the SW structure to the breakdown currents may explain why the SW structures have reached higher maximum gradients than TW structures.

REFERENCES

1. G.A. Loew and J.W. Wang, SLAC-PUB-4647 (1988).
2. G.A. Loew and J.W. Wang, XIVth Int. Symp. on Disch. and Elec. Ins. in Vacuum, Santa Fe, New Mexico, September 16-20, 1990.
3. J.W. Wang *at al*, "High Gradient Tests of SLAC Linear Collider Accelerating Structures," SLAC-PUB-6617 (1994).
4. C. Adolphsen *at al*, "RF Processing of X-Band Accelerator Structures at the NLCTA," LINAC2000, August, 2000, Monterey, Ca.
5. C. Adolphsen *at al*, Paper ROAA003, PAC2001, June 18-22, 2001, Chicago, Illinois.
6. http://www.mrcwdc.com/Magic/

Active and Passive RF Components for High-Power Systems

Sami G. Tantawi and Christopher D. Nantista

Stanford Linear Accelerator Center, 2575 Sand Hill Rd., Menlo Park, CA 94025

Abstract. In recent years, R&D for pulse compression and power distribution systems for the Next Linear Collider has led to the invention of many novel rf components, some of which must handle up to 600 MW of pulsed power at X-band. These include passive waveguide components, active switch designs, and non-reciprocal devices. Among the former is a class of multi-moded, highly efficient rf components based on planar geometries with overmoded rectangular ports. Multi-moding allows us, by means of input phasing, to direct power to different locations through the same waveguide. Planar symmetry allows the height to be increased to improve power handling capacity. Features that invite breakdown, such as coupling slots, irises and H-plane septa, are avoided. This class includes hybrids, directional couplers, an eight-port superhybrid/dual-mode launcher, a mode-selective extractor, mode–preserving bends, a rectangular mode converter, and mode-mixers. We are able to utilize such rectangular waveguide components in systems incorporating low-loss, circular waveguide delay lines by means of specially designed tapers that efficiently transform multiple rectangular waveguide modes into their corresponding circular waveguide modes, specifically TE_{10} and TE_{20} into circular TE_{11} and TE_{01}. These extremely compact tapers can replace well-known mode converters such as the Marié type. Another component, a reflective TE_{01}-TE_{02} mode converter in circular waveguide, allows us to double the delay in reflective or resonant delay lines. Ideas for multi-megawatt active components, such as switches, have also been pursued. Power-handling capacity for these is increased by making them also highly overmoded. We present a design methodology for active rf magnetic components which are suitable for pulse compression systems of future X-band linear colliders. We also present an active switch based on a PIN diode array. This component comprises an array of active elements arranged so that the electric fields are reduced and the power handling capability is increased. Novel designs allow these components to operate in the low-loss circular waveguide TE_{01} mode. We describe the switching elements and circuits.

INTRODUCTION

Because of the requirements of the Next Linear Collider high-power rf systems, passive microwave components have developed significantly during the last few years [1,2]. The power handling capabilities of these components have increased considerably [3]. This has been achieved by increasing the size of these components with respect to the operating wavelength, i.e., by overmoding these components. In particular a class of microwave structures that has complete planar symmetry has been developed. These components carry only TE_{n0} modes. This makes it possible to make all the manipulations in the two dimensional plane. The height of these components can then be increased to reduce the field and allow for high power operation. This class of components is overmoded in both height and width. It allows simultaneous manipulation of multiple TE_{n0} modes; i.e. multimoding. Because of this these components can perform several functions at the same time, resulting in compact and efficient system integration. To make a connection between these modes and circular waveguides needed to transfer rf power over long distances, we've developed a special type of multimoded circular-to-rectangular taper. We present the design of these components and show their application to some rf systems.

Another class of multimoded components that depends on the azimuthal symmetry of circular waveguide carrying the TE_{0n} modes has been developed. This class is used to reduce the length of rf storage lines. We present the design methodology and experimental data for this type of component.

We also extend the spirit of overmoded waveguide structures to active and nonreciprocal devices. We present the theory for a nonreciprocal device that operates in the coaxial circular waveguide mode TE_{01}. This device has the potential of handling tens of megawatts at X-band. It could be used as a circulator or as a switch.

Similar attempts have been made to increase the power handling capabilities of bulk effect semiconductor components. Reports have been made on optically controlled semiconductor switches operating in overmoded waveguide [4]. Here we report the development of a switch made from a PIN diode array operating in an overmoded waveguide carrying the TE_{01} mode. The PIN diode rf switch was invented in the middle of the twentieth century [5,6]. PIN diode switches have wide applications at the low to medium power levels. Various packaged PIN diodes are commercially available. However, due to their small size, packaged PIN diodes cannot be used for switching very high power rf signals. At the end of 1960's, a rectangular waveguide switch [7,8] was developed to handle higher power rf signals than packaged PIN diodes. The window switch was tested up to 100 kW at X-band without problem. Thus, the power handling capability of a semiconductor rf switch has been at the level of 100 kW; this is still low for application to active pulse compression systems for future linear colliders.

A high power device can comprise an array of active elements arranged so that the electric fields are reduced and the power handling capability is increased. In this paper, we will discuss high power rf switches, which consist of the active switching elements. The basic idea is to distribute the power into several elements so that the amount of the power to be handled by each element is reduced, in addition to improving the maximum power of each element by designing it in an over moded structure.

PLANAR COMPONENTS

The idea for planar components has started by the need for a 3dB hybrid capable of handling hundreds of megawatts of rf power at X-band for a SLED-II pulse compression system [9]. A satisfactory device was designed based on the R.H. Dicke's circuit synthesis of a 3dB hybrid [10] (see Fig. 1). Then it was realized that the two interconnecting guides could be combined into one single guide carrying two modes [11] as shown in Figure 1. This resulted in the so-called magic-H hybrid. The device has full planar symmetry, and the height can be adjusted to any value required to reduce surface field and increase power handling capacity.

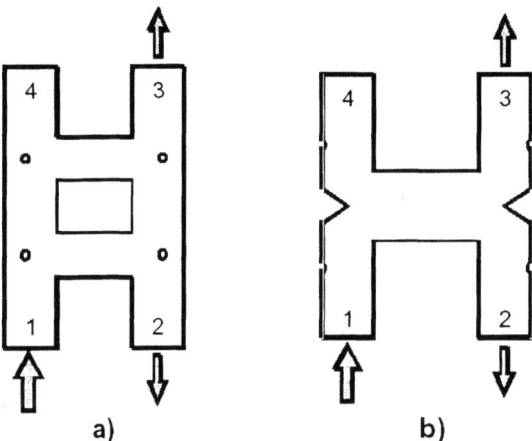

FIGURE 1. Schematic of the h-planar geometries of the a) two-rung ladder and b) "magic H" hybrid designs. Power-flow arrows indicate output ports for the indicated input port.

By putting two of these hybrids together, side-by-side, a dual moded waveguide with similar dimensions as the connecting guide could be produced (see Fig. 2). Adding the remaining parts of the hybrid to this device resulted into the invention of the so-called cross potent superhybrid [12] (see Fig. 2).

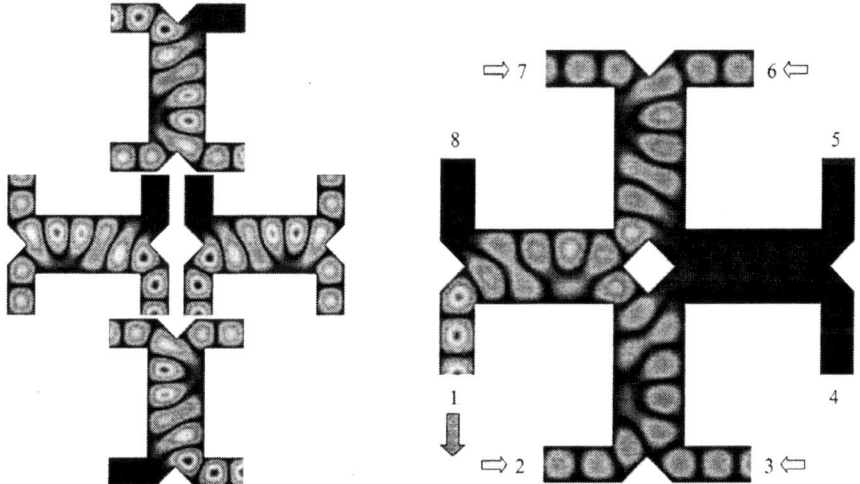

FIGURE 2. Combining four H-plane hybrids resulted in the invention of the Cross Potent Superhybrid

This superhybrid can be used to combine power from four rf sources into any one of four outputs. The choice of the output port depends on the phases of the inputs. The scattering matrix of the device is given by:

$$\mathbf{S}_{cp} = \frac{1}{2}\begin{bmatrix} 0 & 1 & -i & 0 & 0 & -1 & -i & 0 \\ 1 & 0 & 0 & -i & -1 & 0 & 0 & -i \\ -i & 0 & 0 & 1 & -i & 0 & 0 & -1 \\ 0 & -i & 1 & 0 & 0 & -i & -1 & 0 \\ 0 & -1 & -i & 0 & 0 & 1 & -i & 0 \\ -1 & 0 & 0 & -i & 1 & 0 & 0 & -i \\ -i & 0 & 0 & -1 & -i & 0 & 0 & 1 \\ 0 & -i & -1 & 0 & 0 & -i & 1 & 0 \end{bmatrix}$$

One variation on this device is achieved by eliminating two of the output ports in exchange for a single port carrying two modes. Figure 3 shows such a device that can launch either the TE_{10} or the TE_{20} modes in one single port depending on the phases of the input devices. Dealing with two modes at once, TE_{10} and TE_{20}, is possible. For example Figure 4 shows bends that transfer these two modes perfectly at the same time. For the theory for these bends the reader is referred to [13].

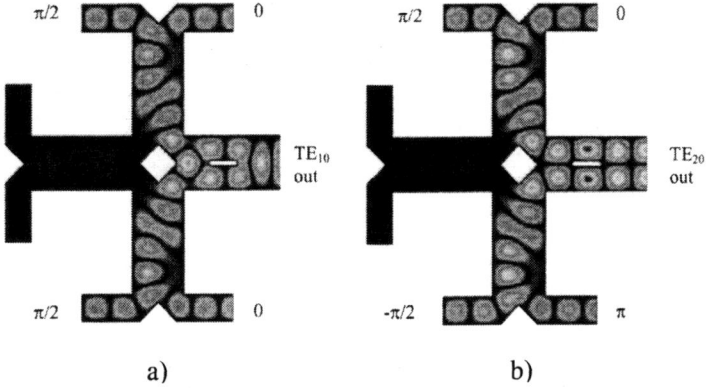

FIGURE 3. Cross Potent Launcher with simulated electric field plots illustrating launching a) TE_{10} and b) TE_{20} in the right overmoded rectangular port with the indicated relative phases for four equal amplitude inputs. Alternate phasings of the inputs send the power to either of the left ports.

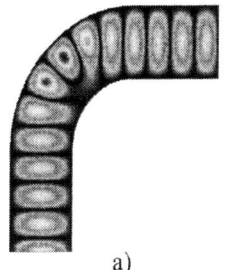
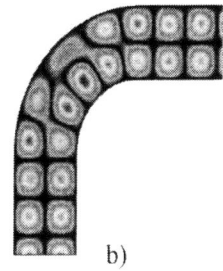

FIGURE 4. Overmoded H-plane bend waveguide with simulated electric field plots illustrating a) TE_{10} mode transmission and b) TE_{20} mode transmission.

One can also mix those two modes by making an H-plane bend in these planar waveguides (see Fig. 5). Using this and the center part of the cross potent superhybrid, one can make a device that separates these modes into different waveguides (see Fig. 5). Indeed, these designs are not unique. For example, another design that can separate these two modes is shown in Figure 6.

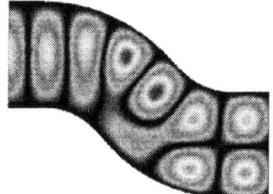

Jog Converter with HFSS simulated electric fields illustrating conversion from TE_{10} to TE_{20} (left to right) or from TE_{20} to TE_{10} (right to left).

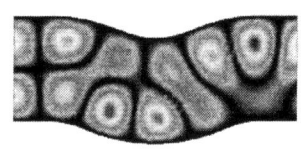

Mode Mixer with HFSS simulated electric fields illustrating conversion from TE_{20} to an equal mixture of TE_{10} and TE_{20}.

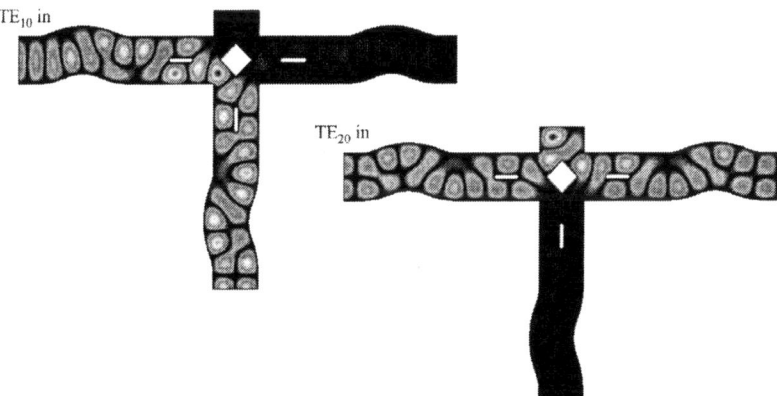

FIGURE 5. Mode mixers and some of their use in mode separators or mode extractors

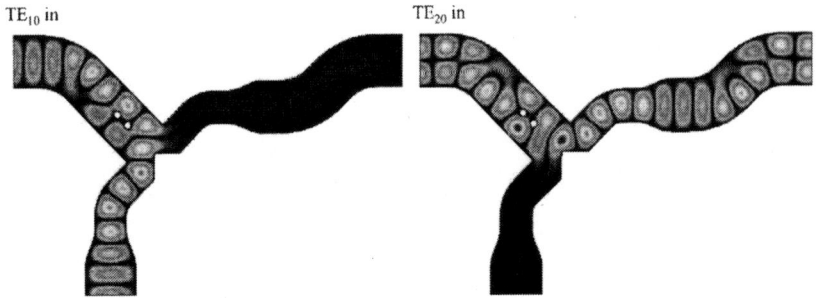

FIGURE 6. Mode-Selective Extractor

To connect these devices with circular waveguides, which is being used for low-loss energy transfer and storage, we use a special type of circular-to-rectangular taper [14]. This taper converts the rectangular guide TE_{10} mode into the TE_{11} mode of circular guide and the rectangular TE_{20} into the low-loss TE_{01} mode of the circular guide (see Fig. 7).

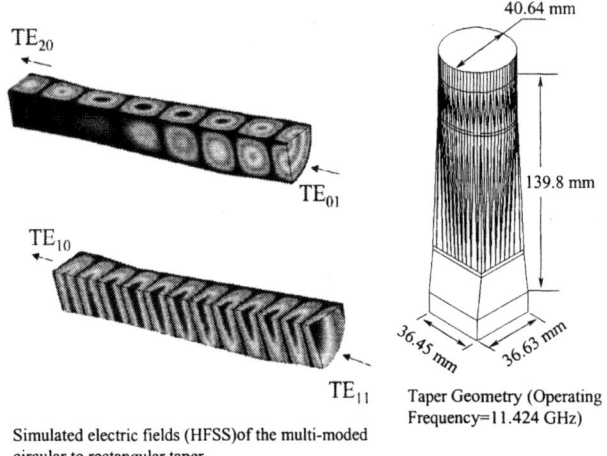

Simulated electric fields (HFSS) of the multi-moded circular to rectangular taper

FIGURE 7. Dual-Moded Rectangular/Circular Converter/Taper

Based on these components, it is possible to design several high power rf pulse compression systems. For example, the Delay Line Distribution System (DLDS) [15] could be greatly simplified using these components (see [13]).

MULTIMODED DELAY LINES

All pulse compression systems considered for the NLC use very long runs of low loss overmoded circular waveguide [16]. The SLED-II system [9] is of particular interest because it minimizes these runs. Yet, even an rf pulse compression system based on SLED-II may use hundreds of kilometers of waveguide for the full installation.

Here we show a method for reducing these long runs of waveguides by making them multimoded. Consider the delay line shown in Figure 8.

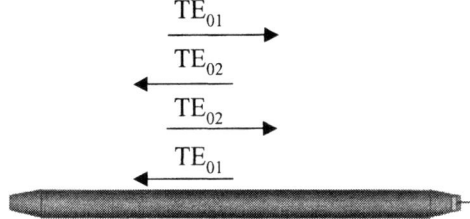

FIGURE 8. Dual-moded delay line

The rf signal is injected into the delay line waveguide in the TE_{01} mode. This is the only azimuthally symmetric TE mode supported at the input port. The waveguide is then tapered up to a diameter that supports several TE_{0n} modes. The TE_{01} mode travels all the way to the end of the delay line and then gets reflected and converted into the TE_{02} mode. The TE_{02} mode travels back to the beginning of this line and, since the input of the line cuts off this mode, gets reflected. If the input taper is designed carefully, TE_{02} can be reflected perfectly. Then, because of reciprocity, the TE_{02} wave gets converted back to TE_{01} at the end of the line. This mode then travels back and exits the line. The total delay in the delay line is twice that seen by a single moded line. Hence, one can cut the length of delay line by a factor of two.

The End Mode Converter

The mode converter at the end of the delay line is shown in Figure 9. It is basically a step in the circular waveguide. If the big waveguide supports only the TE_{01} and the TE_{02} mode among all TE_{0n} modes and the small waveguide supports only the TE_{01} mode, then the device could be viewed as a three-port network. One can choose the diameter of the small guide such that the couplings between each mode in the large guide and the single mode in the small guide are equal. In this case, it is a symmetrical

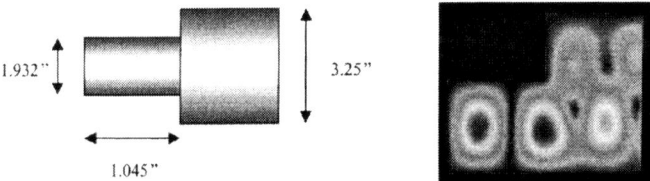

FIGURE 9. TE_{01}-TE_{02} reflective mode converter. The dimensions shown are for an operating frequency of 11.424 GHz. The field pattern shown from finite element simulations predicts a peak electric field of 26.6 MV/m for 300 MW of input power.

three-port network. A theory for such a device is presented in [17]. It shows that there exists a position for placing a short circuit in the middle arm of this three-port network (the small guide in this case) that makes it possible to transfer the power perfectly between the remaining two arms, or in this case between the TE_{01} and the TE_{02} modes in the large guide.

The only step left in the design of this end mode converter is a careful taper design that reduces the diameter of the delay line into the diameter of a waveguide that can support only TE_{01} and TE_{02} modes. The taper needs to transfer both modes perfectly.

Experimental Results

Figure 10a shows the delay through a 75 ns delay line with a short circuit at the end for a total delay of 150 ns. Figure 10b shows the delay after placing the mode converter at the end of this line. The delay was doubled at the expense of increased loss. The loss can be brought back down by using larger diameter waveguide for the delay line.

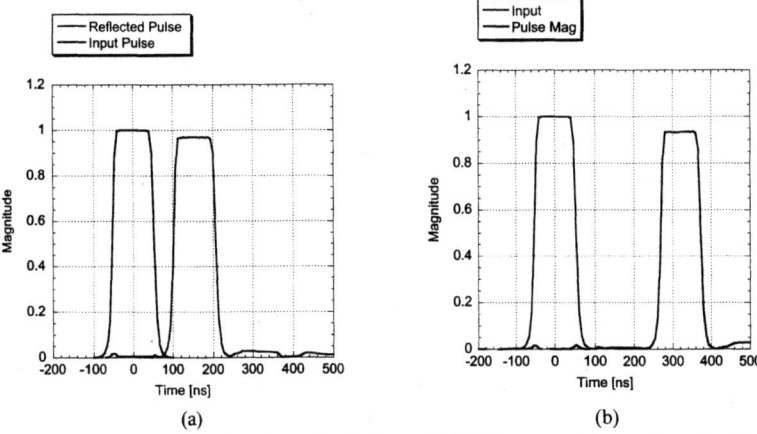

FIGURE 10. (a) Measured delay through 75 feet of WC475 waveguide terminated with a flat plate. The round trip delay time is 154 ns. (b) Measured delay through 75 feet of WC475 waveguide terminated with the TE_{01}-TE_{02} mode converter. The round trip delay time is 320 ns. The operating frequency is 11.424 GHz.

ACTIVE SEMICONDUCTOR DEVICES

A new active window, PIN/NIP diode array active window, which is operated at TE_{01} mode in the circular waveguide was proposed and developed [18]. Our active window is designed and built to avoid the difficulties of the TE_{10} mode rectangular waveguide window switch and to handle X-band rf signals at multi-megawatt levels. This is achieved by
- using an overmoded structure, thus increasing the cross sectional area and reducing the power density.

- using the TE_{01} mode in circular waveguide, which has no electric field at the waveguide wall, thus avoiding edge effects, for a more robust design.

In this section, we describe the design of our PIN/NIP diode array active window.

The Silicon Window

The conceptual view of our active window is illustrated in Figure 11. The base material of the window is highly pure silicon. This window is located in a circular waveguide, which is operated at the TE_{01} mode. As shown in the figure, the PIN/NIP diode structure is a set of radial lines. Each line is heavily doped by P-type and N-type impurities on the topside and backside surfaces, respectively. Each diode line is covered by a metal line, which supplies bias voltage and current to the diode line. The TE_{01} electric field orientation is also indicated in Figure 11. All electric field lines are in the azimuthal direction; i.e. the diode lines are perpendicular to the electric field of the rf signal. This means that the reflection caused by the diode structure and the metal lines is very small when the active window is reverse or zero biased; the rf signal only sees the dielectric contribution of the bulk silicon material. This is the *off* status of the active window. To minimize the reflection from the lines, the *coverage factor*, which is defined as the ratio of the area of the diode structure to that of the whole active region of the window, is chosen to be 10%. The P lines on the front and the N lines on the back are arranged to alternate with each other (see Figure 11). A side-cut view of the active window illustrates the idea (see Figure 12). When forward bias is applied, a massive number of injected carriers goes across the I region. Since the P and N lines are alternating, the injected carriers fill the I region, and the incident rf signal is reflected. This is the *on* status of the active window. The thickness of the active window, which is the same as the I region width in this design, must be small enough so that the carriers injected from the heavily doped P and N lines by the forward bias can diffuse through the high receptivity I region to the heavily doped lines on the other side. This is a very important point to achieve good rf isolation at the *on* status. The carrier lifetime in the high resistivity silicon material is closely related to the diffusion length of the carriers. It is given by $L_D = (\tau D_{AP})^{1/2}$, where L_D is the diffusion length of the carriers, τ is the carrier lifetime in the silicon material, and D_{AP} is the diffusion coefficient. The diffusion coefficient in silicon is $D_{AP} = 15.6$ cm^2/s, giving $L_D \approx 40(\tau(\mu sec))^{1/2} \mu m$. Hence, the base material of the active window must be very pure silicon and must have long carrier lifetime to achieve good conductivity modulation with forward bias. While this puts an upper limit on the window thickness, there is a lower limit for this thickness. If the window is much thinner than the skin depth, good isolation of the rf signal would not be obtained. If we assume the carrier density to be 10^{17} cm^{-3} in the I region with forward bias, then the skin depth δ_s can be calculated with knowledge of the carrier motility in silicon. At a frequency of 11.424 GHz, $\delta_s \approx 100$ μm.

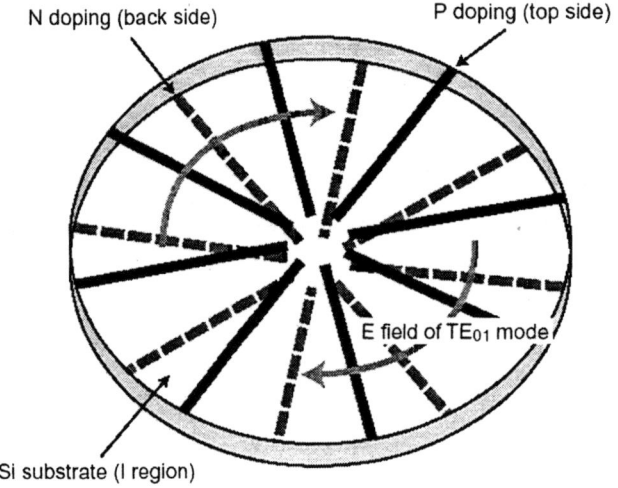

FIGURE 11. Conceptual view of PIN/NIP diode active window.

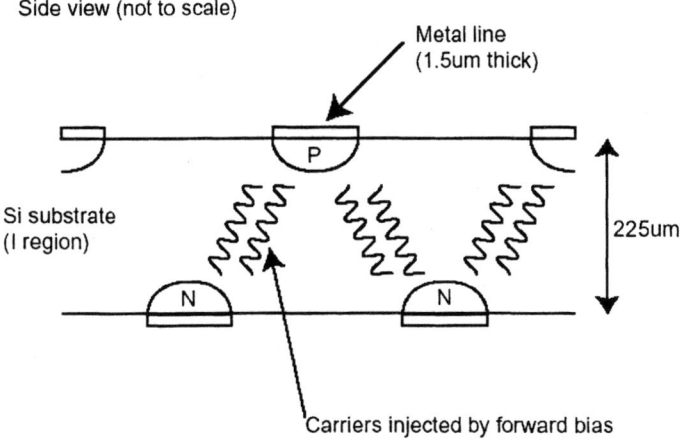

FIGURE 12. Side view of PIN/NIP diode active window.

In the actual design of our prototype active window, the diode structure consists of 400 radial lines each on the front and back. The width of the radial lines is tapered from 25 μm near the waveguide surface to 2 μm near center of the window. The line width and the number of lines are chosen so that the coverage factor is 10%. The thickness of the window is 225 μm. If the carrier density achieved $10^{17} cm^{-3}$, this thickness is more than the skin depth of the intrinsic region for X-band rf signals.

At this carrier density, the surface resistance $R_s \approx 4.8$ ohms at our operating frequency of 11.424 GHz. The diameter of active region and waveguide is 1.299 inches, so that the waveguide impedance is 4.16 times the impedance of vacuum. Hence, the loss dissipated in the window when the window is *on* is given by

$$L_0 \equiv P_l(total)/P_{in} = 4\frac{R_s}{Z_g} \approx 1.23\%.$$

RF Structure and Window Support

The rf structure to support the window is shown in Figure 13. It consists of two aluminum circular waveguides with steps, a ceramic ring, and metal springs. The active window is located between the two waveguides. The ceramic ring fixes the active window at the design position.

The TE_{01} mode rf signal is launched by a compact wrap-around mode converter (see [3]). The diameter of the waveguides changes from 1.5 inches to 1.3 inches at the center by steps designed so that the whole rf structure is matched *without* the window. Since silicon has a large dielectric constant, the impedance mismatch at the surface of the active window is large, causing non-negligible reflection. Because we are interested in the pure characteristics of the active window, the structure was designed without matching sections to compensate for the mismatch at the window surfaces.

The two waveguides are DC separated. The ceramic ring works as the DC voltage gap. There are metal springs between the active window and the waveguides. The waveguides are connected to a biasing circuit. The biasing voltage is supplied to the active window through the waveguides and the springs, and the waveguides are DC isolated from the TE_{01} mode converters by mylar insulators.

There is a gap between the two waveguides, but no choke structure. Since the surface currents of TE_{01} mode in circular waveguides are azimuthal, the gap does not cut any surface currents and there is no rf leakage through the gap. Indeed, this a big advantage of this structure over TE_{10} mode rectangular window switches; we can avoid the complex choke geometry necessary in the rf structures of this later type.

Finally, vacuum seals necessary for high power operation under vacuum are made with Viton rubber gaskets for DC isolation.

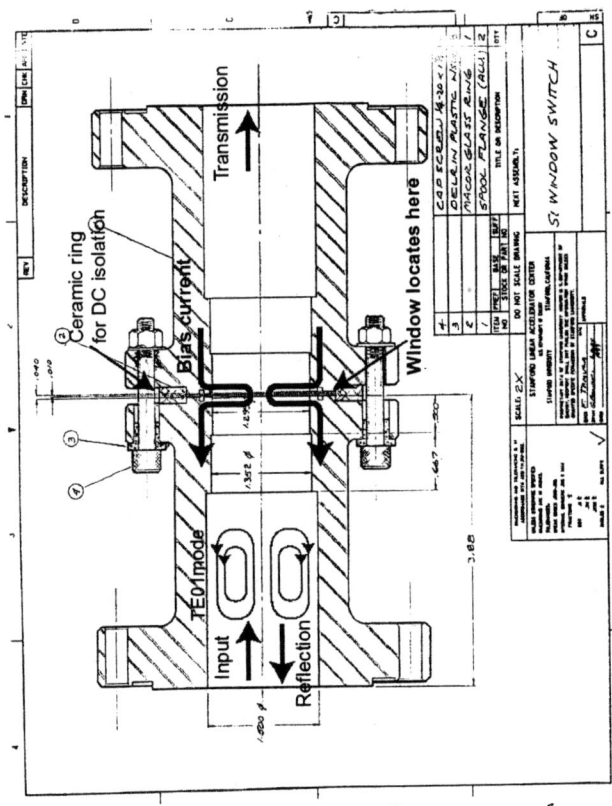

FIGURE 13. Active window and rf structure.

Experimental Results

The high power experiment was performed with an X-band klystron (XL-2) and a SLED-II rf pulse compression system at SLAC. A high power rf signal of 900 ns pulse duration at 11.424 GHz was generated by the klystron, compressed by the SLED-II system, and fed to the active waveguide window. The SLED-II system allowed us to operate the klystron at a relatively low power level and acted as a buffer between the klystron and the unmatched window. The power level output to the active window was up to 15 MW. For our experiments, this was high enough. The compressed rf pulse duration was 150ns. The repetition rate was limited to 5 Hz, since the active window did not have cooling.

The incident, reflected, and transmitted rf signals were measured by power meters and rf diode detectors through directional couplers. The TE_{01} mode converters have view-ports to watch the surfaces of the window so that the video camera could detect flashes of light if arcs occur on the surface of the active window.

Two types of silicon windows were prepared for this high power experiment. First one is an active window, which has a full PIN/NIP diode array structure on both the front and back. This active window is the 10^{17} cm^{-3} doping density version. The resistivity of the base material is 5000 ohm-cm. The window thickness is 225 µm. The second one had only the metal line structure on one side and no doping line structure; the other side had no structure (*metal-only version*). This version of the window was prepared for investigating the breakdown properties of the thin metal structure. The resistivity of its base material is 1000 ohm-cm and its thickness is 315 µm.

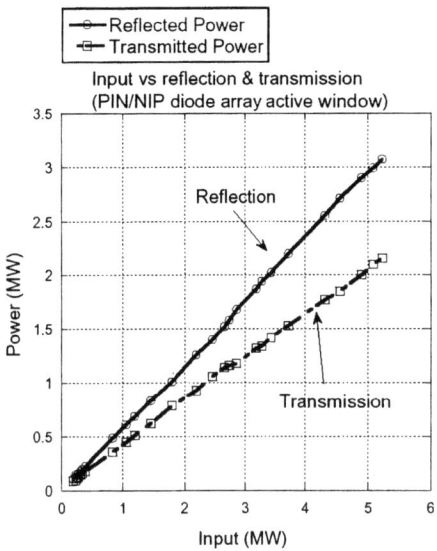

FIGURE 14. Input versus transmission and reflection power.

In Figure 14, the reflected and transmitted powers from the active window with full diode structure are plotted. As shown in the figure, the reflected and transmitted powers were proportional to the input power. Hence, the reflection coefficient did not change and the loss dissipated into the active window did not increase with increasing the input power up to 5.2MW. This means that avalanche breakdowns did not occur at this power level. If the avalanche breakdown occurred, there would be copious carriers in the intrinsic region, and the reflection coefficient would have changed.

However, arcing started at an input power level of around 4 MW. When arcing occurred, the vacuum in the system went from 10^{-8} Torr at normal operation to above 10^{-5} Torr, and interlocks stopped the klystron. After the onset of arcing, the reflected power increased and the transmitted power slowly decreased.

This arcing was very vacuum dependent. We had to process slowly from the lower power levels after the trips. Flashing lights were observed by the video camera from both sides when the arcing occurred. These characteristics, vacuum dependency and flashing lights, are typical characteristics of the vacuum breakdowns, not of avalanche breakdowns.

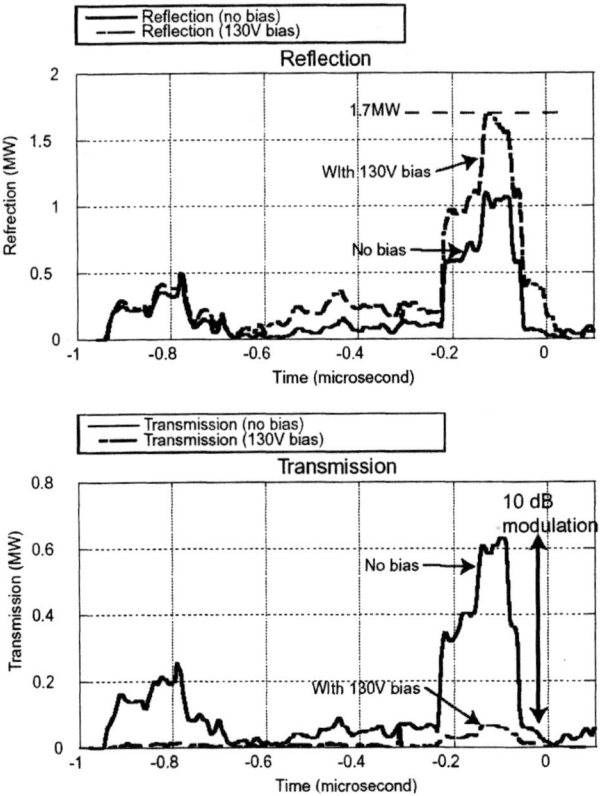

FIGURE 15. Waveform with zero bias and 130V forward bias.

After achieving the maximum input power of 5.2 MW, the arcing occurred more and more frequently, and also at lower power levels. We could not raise the input power further. Thus, the high power operation was limited by arcing. The maximum field at the surface of the window is calculated as 3.8 MV/m at 5.2 MW.

The waveforms of the transmitted and reflected rf signal with and without the forward bias voltage are shown in Figure 15. With the forward bias voltage at 130 volts, the reflection increased and the transmission decreased. The reflected rf signal was modulated from 1.1 MW to 1.68 MW, and the transmitted rf signal was modulated from 0.63 MW to 60 kW. A 10 dB of transmission modulation was thus achieved. With the forward bias, no arcing on the surface of the active window was observed, which is reasonable because the active window is a conductor, and the electric field of rf signal is minimum near the window surface.

ACTIVE MAGNETIC DEVICES

The implementation of a circulator or a switch can be achieved using a two-port nonreciprocal network plus a 3dB hybrid and a splitter (see Fig. 16). The simplest implementation of the nonreciprocal element in an overmoded waveguide using the TE_{01} mode is shown in Figure 17. This implementation also depends on the so-called wrap-around mode converter [3]. In this system the mode converter launches the TE_{01} mode, which has both axial and radial magnetic fields, in a coaxial structure. At the ends of the structure, the coaxial guide becomes narrow and only the coaxial TEM mode can propagate (see Fig. 2). A pulsed current signal could be launched from that narrow port (port-A in Fig. 2) This pulse would have only azimuthal magnetice field. This field is used to bias a piece of garnet wrapped around the center conductor of the coaxial structure. This structure has several advantges for handling high power rf signals:

1- Operating in an overmoded waveguide, with a large cross sectional area for a given wavelength, should give the device a high power handling capacity.
2- All rf electric field lines are parallel to the interface between the garnet material and vacuum (see the theory section below).
3- The center conductor could be used to cool the garnet material; it could be made as a tube with water flowing through the center.

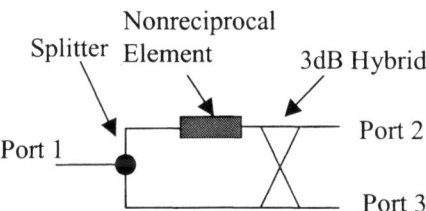

FIGURE 16. This three-port network will work as a circulator if the phase shift through the two-port nonreciprocal element differs by 180 degrees for different propagation direction. The system could work as a switch if one can control the phase shift through the element.

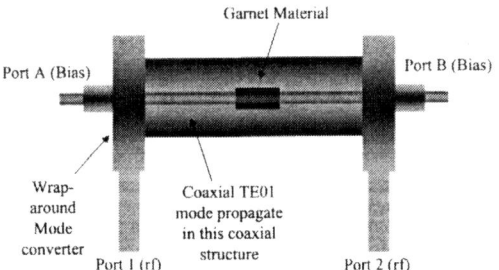

FIGURE 17. Nonreciprocal two-port device employing the TE_{01} mode.

Several variations on this device are possible. An elegant implementation of the system shown schematically in Figure 16 is illustrated in Figure 18. In this geometry the splitter is realized by dividing the power between the TE_{01} and the TE_{02} modes. These two modes interact with the garnet coated section in different manners. To implement the circulator, one could design the system to make the phase difference between these two modes in the forward direction differ from that of the backward direction by π. To implement a switch, one could control that phase difference by varying the current in the center conductor.

For a proof of principle numerical experiment we considered the properties of calcium vanadium doped garnet [19]. We chose this material because of its narrow line width. The calculated values for μ and k of its permeability tensor are shown in Figures 5 and 6. The operating frequency used to generate these curves is 11.424 GHz. The dielectric constant of this material is about 14. The biasing magnetic field was chosen so that the material would operate below the resonance frequency.

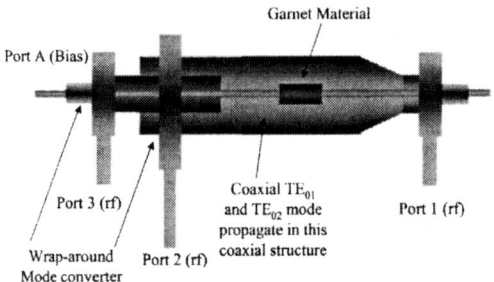

FIGURE 18. Implementation of the overmoded circulator/switch.

Figure 19 shows the required garnet thickness and length for a π phase difference between forward and backward TE_{01} mode waves. For the optimum thickness, the rf losses are less than one percent. This is a manageable loss level whose heat can be removed by cooling through the center conductor.

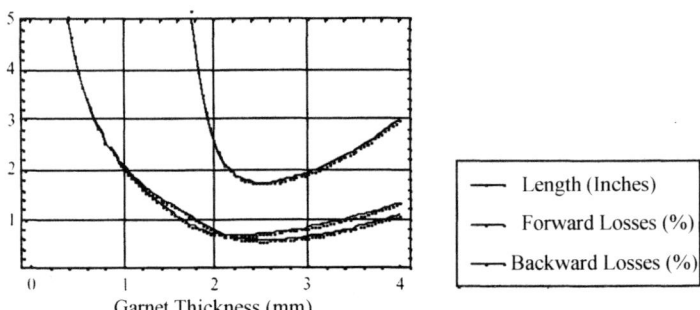

FIGURE 19. The required dimensions of the garnet material for a π phase shift between forward and backward waves for the TE_{01}.

CONCLUSIONS

In this paper, we have presented an overview of some developments made over the past few years in the area of high power microwave components. This work has been motivated by the requirements of an rf system for a next generation electron-positron linear collider, which must handle X-band power levels up to hundreds of megawatts. These components are thus all overmoded to ameliorate breakdown problems associated with high fields.

They include several passive waveguide devices that make use of multi-moding for directing power through phasing of combined sources. These take advantage of planar symmetry for ease of mode manipulation and height-independence of the designs. They also include mode converters between rectangular waveguide modes, between rectangular and circular waveguide modes, and between circular waveguide modes. The latter device can be used to double the storage time in a reflective delay line.

Some novel active components, activated by pulsed voltage or current, have also been described. These include a PIN diode array on a silicon wafer operating as an rf window and a garnet based nonreciprocal waveguide structure. Some experimental results for the former at multi-megawatt operation have been presented. Both take advantage of the TE_{01} mode field pattern in circular waveguide. Either of these devices, although they require further development, have the potential of providing the basis for a high power microwave switch.

ACKNOWLEDGMENTS

The authors would like to thank Professors Ronald Ruth, Perry Wilson, Norman Kroll and Roger Miller for many useful discussions. We would like to thank Gordon Bowden for his help in constructing the rf structure. The semiconductor switch work was done in collaboration with Fumihiko Tamura. We are also grateful for the help and effort of Chuck Yoneda and the vacuum group at the Klystron Department at SLAC.

This work was supported the US Department of Energy contract DEAC03-76SF00515.

REFERENCES

1. Sami G. Tantawi, "New Development in RF Pulse Compression," Proc. of the XX International Linac Conference, Monterey, California, USA, August 21 - 25, 2000, p. 673-677, and the references cited therein.
2. Christopher D. Nantista and Sami G. Tantawi, "A Planar Rectangular Waveguide Launcher and Extractor for a Dual-Moded RF Power Distribution System," Proc. of the XX International Linac Conference, Monterey, California, USA, August 21 - 25, 2000.

3. S.G. Tantawi, et al., "The Generation of 400-MW RF Pulses at X Band Using Resonant Delay Lines," *IEEE Trans. Microwave Theory Tech.*, vol. 47, no. 12, pp. 2539-2546, Dec. 1999; SLAC-PUB-8074.
4. Sami G. Tantawi, Ronald D. Ruth, Arnold E. Vlieks, and Max Zolotorev, "Active high-power RF pulse compression using optically switched resonant delay lines," IEEE Transactions on Microwave Theory and Techniques, MTT-45(8), August 1997, p. 1486–1492.
5. S. M. Sze, "Physics of Semiconductor Devices," John Wiley & Sons, second edition, 1981.
6. Joseph F. White, "Microwave Semiconductor Engineering," Van Nostrand Reinhold Company, New York, 1982.
7. Kenneth E. Mortenson, Jose M. Borrego, Paul Bakeman, E. Jr., and Ronald J. Gutmann, "Microwave silicon windows for high-power broad-band switching applications," IEEE Journal of Solid-State Circuits, SC-4(6), December 1969, p. 413–421.
8. Kenneth E. Mortenson, Albert L. Armstrong, Jose M. Borrego, and Joseph F. White, "A Review of Bulk Semiconductor Microwave Control Components," Proceedings of the IEEE, 59(8), August 1971, p. 1191–1200.
9. P. B. Wilson, Z. D. Farkas, and R. D. Ruth, "SLED II: A New Method of RF Pulse Compression," Linear Accl. Conf., Albuquerque, NM, September 1990; SLAC-PUB-5330.
10. Montgomery, Dicke, and Purcell, <u>Principles of Microwave Circuits</u>, Rad. Lab. Series, 1948, p. 451.
11. C.D. Nantista, W.R. Fowkes, N.M. Kroll, and S.G. Tantawi, "Planar Waveguide Hybrids for Very High Power RF," proc. Of the 1999 IEEE Particle Accelerator Conference, New York, N.Y., March 29-April 2, 1999.
12. Christopher D. Nantista and Sami G. Tantawi, "A Compact, Planar, Eight-Port Waveguide Power Divider/Combiner: The Cross Potent Superhybrid," IEEE Microwave Guided Wave Lett., vol. 10, no. 12, pp. 520-522, December 2000; SLAC-PUB-8771.
13. Christopher D. Nantista and Sami G. Tantawi, "Multi-Moded Passive RF Pulse Compression Development at SLAC," proc. of Advanced Accelerator Concepts Workshop, Santa Fe, NM, June 10-16, 2000.
14. S.G. Tantawi, N.M. Kroll, and K. Fant, "RF Components Using Over-moded Rectangular Waveguides for the Next Linear Collider Multi-Moded Delay Line RF Distribution System," Proc. Of The IEEE Particle Accelerator Conference, New York City, March 29th - April 2nd, 1999, p. 1435-1437.
15. H. Mizuno, Y. Otake, "A New Rf Power Distribution System For X Band Linac Equivalent To An Rf Pulse Compression Scheme Of Factor 2**N," 17th International Linac Conference (LINAC94), Tsukuba, Japan, Aug 21 - 26, 1994.
16. S.G. Tantawi, R.D. Ruth, and P.B. Wilson, "A Comparison Between Pulse Compression Options for NLC," proc. of IEEE Particle Accelerator Conference (PAC 99), New York, 29 Mar - 2 Apr 1999. Published in *New York 1999, Particle accelerator, vol. 1* 423-425.
17. Sami G. Tantawi and Mikhail I. Petelin, "The Design and Analysis of Multi-Megawatt Distributed Single Pole Double Throw (SPDT) Microwave Switches," in IEEE MTT-S Digest, 1998, pp. 1153–1156.
18. Fumihiko Tamura and Sami G. Tantawi, "Development of High Power X-band Semiconductor RF Switches for Pulse Compression Systems of Future Linear Colliders," proceedings of the XX International Linac Conference, Monterey, California, USA, August 21 - 25, 2000.
19. http://www.trans-techinc.com/catalog/pdfs/pg1-4&5.pdf

Power Supply, Energy Storage Line, and Grid Pulsers for High Voltage Gridded Klystrons

Roland F. Koontz

SLAC 2575 Sand Hill Road MS 33
Menlo Park, CA 94025
rfkap@slac.stanford.edu

Abstract. Designs for high power, gridded klystrons are being considered for driving accelerators. These designs have high voltage DC on the klystron cathodes, with the klystron current being turned on and off with a much lower voltage grid drive pulse. Such a klystron eliminates the need for a high power pulse modulator. The modulator is replaced by a high voltage energy storage line, an RF switching line charging supply, and a small electronics package consisting of a DC grid bias supply, a fast rise and fall time grid pulser, and a klystron cathode heater power supply. This paper outlines some of the design details of such a gridded klystron support system including specifications for the energy storage cable, and the fast grid pulse driver. Such a system can be very compact and reliable with low initial cost, and excellent operating efficiency.

BACKGROUND

Various linear collider schemes require large amounts of microwave power that are delivered by large quantities of klystrons. R&D effort centers on developing these klystrons, and the power sources that drive them. One scheme that has the potential to produce economical microwave power at reasonable capital cost is the sheet beam gridded klystron and its supporting high voltage energy storage cable, grid pulser, and DC high voltage power supply. A Phase I SBIR award was made in 2001 to develop a sheet beam gridded gun. Sheet Beam RF structures that can use such an electron beam source are the subject of R&D at SLAC. This paper describes the initial design concept of the supporting power source for this class of klystrons. A 2002 SBIR solicitation for the cable and supporting electronics is currently published in the DOE SBIR solicitation listing.

SYSTEM DESCRIPTION

The voltage and current of a dual sheet beam electron gun to power an X band klystron is subject to design optimization as the development proceeds, but for this example we will consider a dual sheet beam gridded electron gun with an initial triode stage and a post acceleration stage. The total DC cathode voltage on the gun is 420 kV with the initial triode stage being operated at 150 kV, and the post acceleration stage at 270 kV. Total beam current from the dual cathode gun is 560 amps during a pulse of 600 nanoseconds. The voltage of the intermediate electrode separating the triode stage from the post acceleration stage is maintained through a high impedance feed, and functions as an arc suppressor. In the triode stage, the positive grid drive pulse required is plus 5 kV operating from a negative bias voltage of minus 1 kV. Klystron cathode heater voltage is 14 volts DC, and current is 40 amps. As the electron gun design is optimized, the beam voltage could climb as high as 500 kV.

The power supply support system envisioned for this klystron consists of a multiple concentric conductor high voltage cable that provides primary pulse energy storage for the klystron electron gun when pulsed while also connecting the klystron to the remote grid pulser and power supply system. This remote system consists of a primary switching high voltage DC power supply with intermediate tap, a 5 kV, 600 nanosecond grid pulser, a 1 kV DC bias supply, and a 1 kW klystron cathode heater supply all floating on the top of the DC supply. A diagram of this system is shown in Fig. 1.

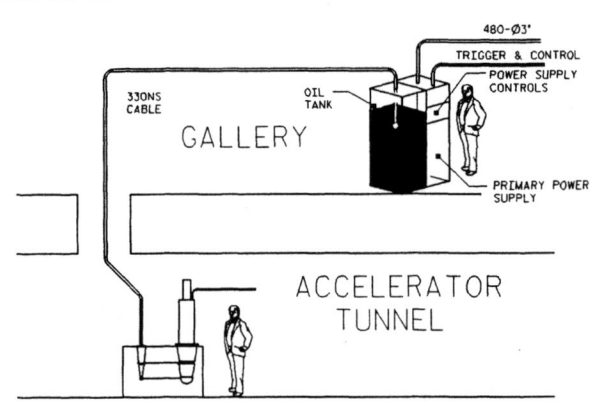

FIGURE 1. Gridded Klystron Power System

Part 1 the Energy Storage Cable

The power scheme proposed uses a high voltage, multiply concentric conductor cable to store the energy delivered during the short, 600 nanosecond klystron cathode pulses. The pulse repetition frequency of these pulses is 120 Hz. The dynamic impedance of the klystron during the pulse is on the order of 750 ohms. A typical cable impedance for this sort of cable design is 35 ohms. Thus, if the cable is initially charged to 5% over required cathode voltage, when the grid is pulsed and the cathode delivers 560 amps, the cable voltage on the load end will drop to the required cathode voltage, and this voltage will be maintained until the wave-front launched on the cable as the result of the grid switched cathode current travels to the other end of the cable and returns to the load end. At this time, the grid turns off the cathode current, which cancels the returning wave. This dictates that the cable must have an electrical length of exactly half the cathode pulse width, 300 nanoseconds. The cable is then recharged slowly during the interpulse period. The cable must have good DC high voltage standoff characteristics, while also having very low loss and dispersion functions for the traveling waves.

As a simple example, wave propagation on a standard coaxial line of 50 ohms when charged and then switched into a 50 ohm matched load is shown in Fig. 2. In this case, the voltage the line is charged to drops by 50% when the switch is closed, and all of the energy stored in the line empties out in a 600 nsec flat pulse.

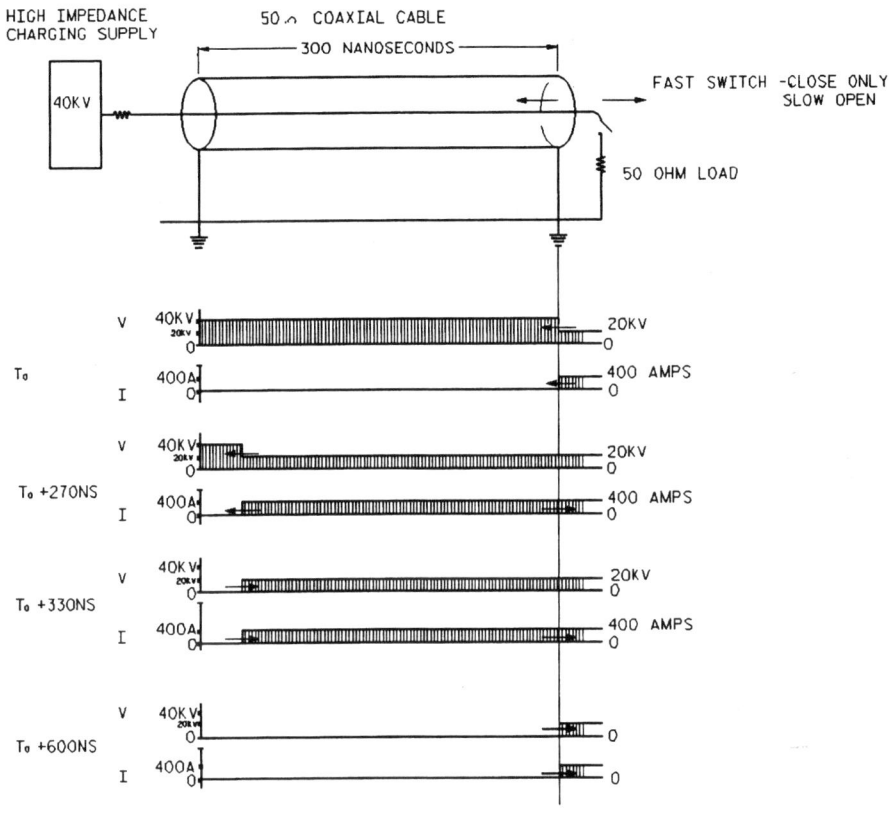

PULSE FORMATION WITH FULL DISCHARGE OF COAX LINE

FIGURE 2. Matched discharge from a 50 ohm line

A similar situation prevails in a line which is much lower impedance than the load, but where the load can be switched both on and off with fast rise and fall times, and a 600 nsec on duration. The wave propagation on this line is shown in Fig. 3. In this case, some of the potential capacitive energy stored in the line is transformed into a pair of traveling waves, one flowing forward into the load (the klystron) and a second wave moving back up the line. This backward traveling wave is reflected off of the input end of the open circuited line. If, when the front of this wave reaches the klystron end of the line, the grid of the klystron turns off the klystron beam current, all waves cancel, and the line is left with no traveling waves, but with potential energy reduced by twice the initial line voltage drop which occurred when the klystron was initially turned on. This used potential energy is then replenished from the charging supply during the inter-pulse period as the line is recharged to its initial potential.

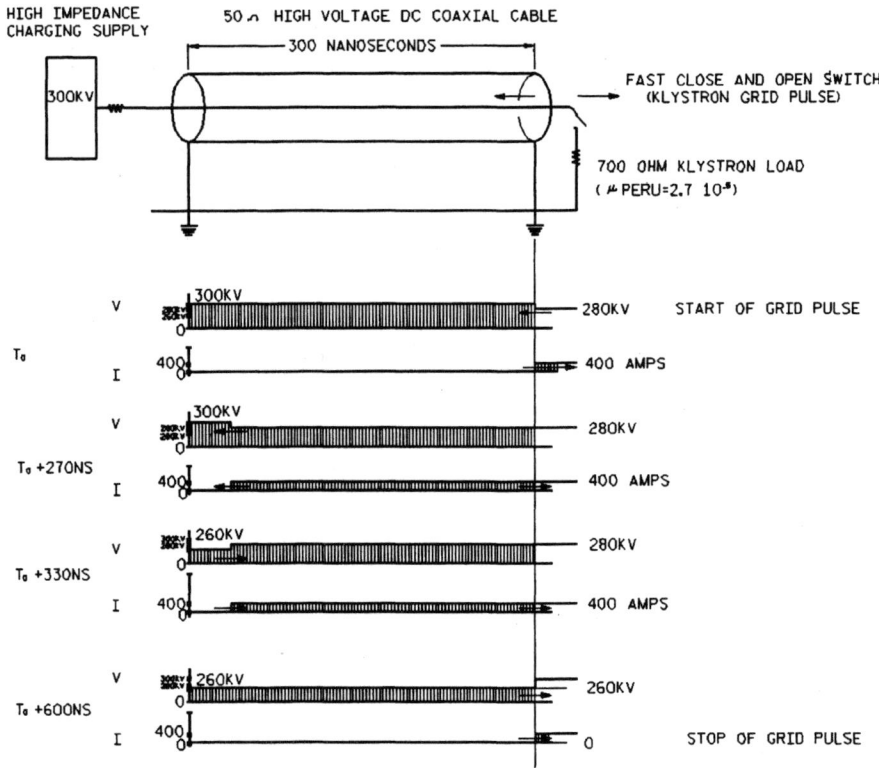

FIGURE 3. Partial discharge of mismatched high voltage line

The line construction consists of multiple concentric conductors as shown in Fig. 4 on the next page.

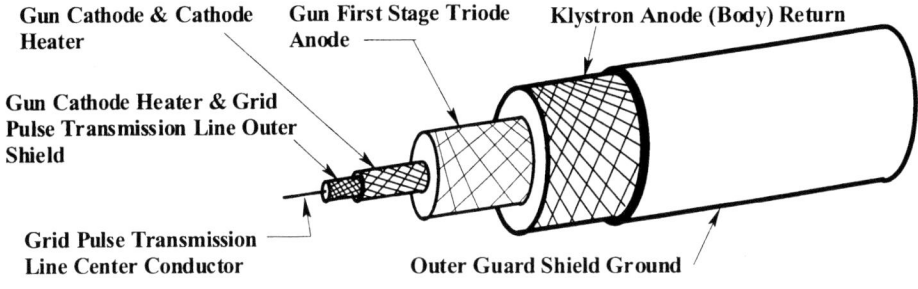

FIGURE 4. Multiply concentric high voltage line

Innermost coax: 50 ohms, low loss and dispersion to carry the grid bias and grid pulse (600 nanosecond pulse, rise and fall times of less than 10 nanoseconds). Outside of coax carries heater current of 40 amps.

Second-Third Shells: Inner shell is the cathode heater connection, outer shell is klystron cathode which is tied to the other leg of cathode heater. Each shell must carry 40 amps.

Third-Fourth Shells: The dielectrics between these two shells hold the primary energy storage of the cable. This shell is maintained at a voltage of 150 kV, and provides the cathode-anode voltage for the klystron triode first stage. A semiconductor layer in the load connector provides the high impedance connection to this intermediate gun electrode. The cathode pulse energy is delivered through this, and the following shell dielectrics.

Fourth-Fifth Shells: The dielectric between these two shells holds the other part of the primary energy storage of the cable at a voltage of 270 kV, and provides the post acceleration voltage for the klystron gun second stage.

Since high voltage DC cables are extensively used in both the power generation and transmission industries, and X-ray facilities as well, characterization and design of the cable needed for this application should be straight forward, and the production costs well within the boundaries of this klystron support system. Fast pulse propagation studies must be done to optimize dielectric materials for both high voltage hold-off and high frequency transmission characteristics.

Part 2 High Voltage Charging Supply, Heater, Bias, and Pulser Floating Deck

The supply side of the energy storage cable is connected to a 450 kV high impedance capacitor charging power supply that incorporates a small floating deck on the high voltage terminal and junction with the energy storage cable. A means of providing about 2 kW to this floating deck must be provided to power the electronics described below. The power supply should provide –450kV to ground on the floating deck terminal, and an intermediate tap at –290 kV for the klystron gun intermediate anode. Capacitive energy storage on the load side of the power supply must be kept to a minimum to limit energy dissipation in an arc, and provide a high impedance to approximate the required open circuit condition of the supply side of the cable during pulse energy discharge. It is anticipated that this power supply design is a fast switching circuit using IGBT's and a voltage multiplier. The average power of the supply is about 20 kW, and should operate from a 480 volt, three phase primary power source. It should incorporate fast fault sensing and fast shutdown capability. The efficiency of this supply directly affects linac operating costs, and should be as high as possible, over 95%.

Part 3 Klystron Cathode Heater, Bias, Grid Pulser

Heater and Bias supply designs are conventional electronics, but must be packaged to mount on the high voltage deck associated with the power supply above. The grid pulser drives the input end of the innermost coax of the high voltage cable (50 ohms). The pulser must supply a flat (1%) 600 nanosecond pulse with rise and fall times of less than 10 nanoseconds each to this line. The output circuit of the pulser must present a 50 ohm source impedance to the cable to absorb reflections from the other end of the energy storage cable that result from the non-linear impedance of the grid-cathode elements. The pulser must also be isolated from the high voltage deck (klystron cathode potential) and float on top of the negative grid bias supply. The pulser should be fiber-optically triggered, with a triggering stability of less than 1 nanosecond.

SUMMARY

The system outline presented above is based on many proposals and experiments described in literature dating back almost fifty years. New advances in klystron design incorporating sheet beams and electron guns with grids now make such systems feasible for a wide variety of uses including accelerator RF systems. Industry already possesses much of the basic technology needed to bring such a klystron support system into existence. With the use of appropriate SBIR funding, a system as described can be implemented to match the ongoing klystron development effort.

Development of an X-band RF Gun at SLAC

A. E. Vlieks*, G. Caryotakis*, W. R. Fowkes*, E. N. Jongewaard*,
E. C. Landahl◊, R. Loewen* and N. C. Luhmann, Jr.◊

*SLAC, 2575 Sand Hill Rd, Menlo Park, CA 94025, USA
◊ 3001 Engineering III, Dept. of Applied-Science
Davis, CA 95616, USA

Abstract. As part of a National Cancer Institute grant to develop a compact source of monoenergetic X-rays via the Compton Effect, we have completed the design of a Laser-driven 5.5 cell RF gun operating at 11.424 GHz. The goal is to develop an RF Gun which can generate a 7 MeV, 0.5nC electron beam with an RMS emittance of >1 π-mm-mR. We have completed simulations of the total beamline, including the RF gun, accelerator structure, focusing quadrupole triplet and electron beam/laser beam Interaction Region using PARMELA. Results of these simulations will be presented, showing that a 60 MeV electron bunch can be focused to an interaction point two meters downstream of the photocathode. We will also present results of RF measurements of the Gun-cold-test model showing the field distribution along the gun axis and the gun resonances. Details of the RF power Source, Emittance-Compensating Solenoid and Laser system will also be presented

INTRODUCTION

One of the proposed methods of identifying and destroying certain types of cancer cells is through the use of tunable monochromatic X-Rays [1]. Heavy elements such as Iodine, Gadolinium or Gold can be introduced near or in cancer cells by attaching them to specially designed molecules, which are selectively absorbed by tumors. By irradiating these cells with X-Rays at the K-shell energy of the heavy element, the absorption cross section for these elements is enhanced relative to the neighboring tissue. If a second X-ray image is taken at an energy below the K-shell energy and subtracted from the first image, the resulting image spotlights the heavy element and. thereby helps to identify the location of the tumor. A higher X-Ray flux can be used to selectively destroy these cells. This technique has already been investigated in large Synchrotron facilities [2].

We are working on the development of a tunable X-ray source using the Compton effect to upshift the energy of a high power laser beam into the X-ray region using a low emittance electron beam as the energy source. In our setup, shown in figure 1, a 7 MeV electron beam is generated by a 5.5 cell X-band RF gun, operated at X-band, and driven by a UV laser. The beam energy is increased up to 60 MeV using a 1.05 m X-band accelerator. Finally it is focused to a minimum spot size at the Interaction Region where it collides, nearly head-on with a high power 800 nm laser beam, and generates X-rays to energies up to > 87 KeV.

The electron beam is then removed from the beamline axis with a spectrometer into a beam dump. A pair of 50 MW klystrons is used as high power RF sources.

One klystron will feed the RF gun while the second will drive the accelerator. Both are driven from a common signal generator to maintain phase coherence.

In this paper we will describe the details of the proposed experimental setup, simulations of the RF gun and RF measurements of a cold-test prototype.

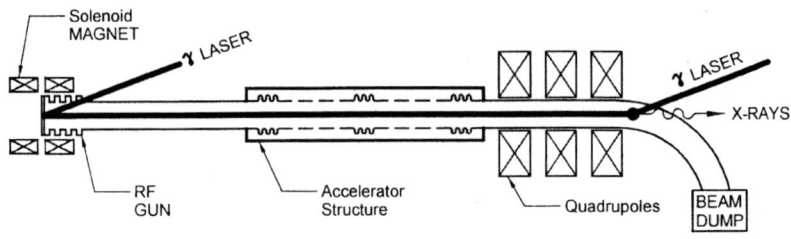

FIGURE 1. Schematic of Monochromatic X-Ray Source

EXPERIMENTAL SETUP

The design parameters of the gun are shown in Table 1. In order to attain an electron beam with energy of approximately 7 MeV with a maximum surface gradient of 200 MV/m we were required to design a gun with 5.5 cells. This number of cells is unique among rf gun designs. In many rf gun designs, performance is often limited by RF breakdown at the boundary where the demountable cathode fits into the gun. In our initial configuration we will not use a demountable cathode but instead use the whole back wall of the half-cell as the cathode. It will be made from OFE copper and machined to a micron-level surface flatness.

A (nominally) 6 KG magnet has been designed using POISSON [3] to serve as the emittance-compensating solenoid. The magnet is designed as two identical pairs

TABLE 1. Gun Design Parameters

Beam Energy at gun exit (for 200MV/m gradient)	7 MeV
Bunch Charge	0.5 nC
Transverse emittance	1 π-mm-mR
Laser spot size at cathode (radius)	0.5mm
Laser temporal length	800 fs
Beam spot size at Interaction Region (radius)	20 microns
Beam energy at Interaction Region	60 MeV

operated so that there is exactly zero magnetic field in the center where the cathode will be positioned. This is shown in figure 2.

The gun laser beam, operating at 266nm, the third harmonic of a Ti:Sapphire laser, is directed towards the cathode at nearly normal angle of incidence after being deflected by a fixed mirror in a diagnostic chamber (located approximately .5 m downstream of the gun). All adjustments to the laser beam are performed external to the vacuum envelope. The current design calls for a dielectric mirror but metallic mirrors will also be investigated. A second mirror, symmetrically located with respect to the beam axis, is designed into the chamber to monitor the (cathode) reflected laser beam for diagnostic purposes.

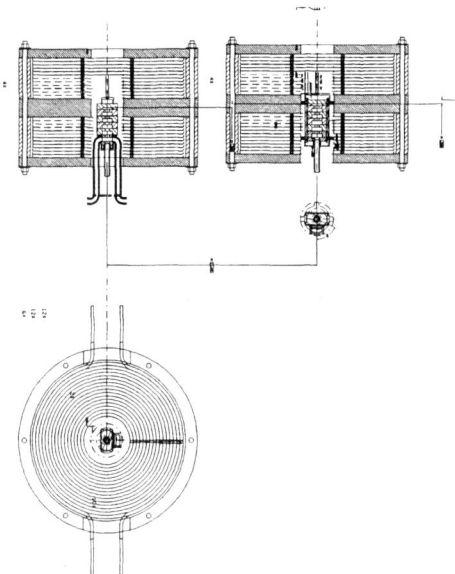

FIGURE 2. Emittance Compensating Solenoid

The chamber is shown in figure 3. Besides serving as a holder for the laser mirrors it also permits the insertion of various diagnostics (e.g. YAG crystal, OTR foils, pepper pot masks etc.) into the beam and contains a view port for observing the beam profile with CCD camera.

A 1.05 m accelerating structure is positioned approximately 67 cm downstream of the cathode. This structure is a $2\pi/3$ traveling wave structure. This structure has been successfully conditioned to approximately 75 MV/m (accelerating gradient) during Next Linear Collider (NLC) breakdown studies. For our purposes we will only need a maximum accelerating gradient of approximately 53 MV/m. The RF power requirement at this gradient is approximately 40 MW.

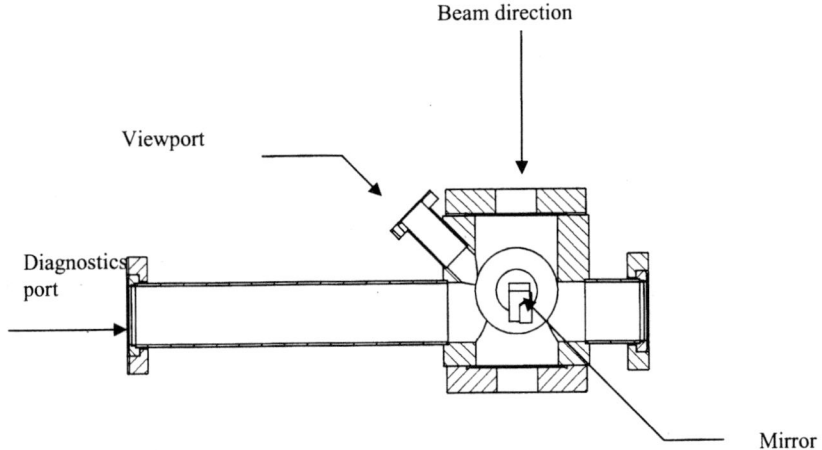

FIGURE 3. Diagnostic Chamber

Downstream of the accelerator is located a triplet of quadrupoles which will focus the beam to a spot size of ~20 microns in the interaction chamber (not shown in fig.1).

SIMULATIONS

The gun cavities were designed to operate at 11.424 GHz in the π-mode using the code SUPERFISH[3]. Results of these simulations, shown in figure 4, indicate that approximately 16 MW of RF power will be required to develop the peak design gradients at the cathode. Power will be fed into the gun through symmetric ports in the last (most down-stream) cell in order effectively remove the dipole asymmetry present in a single feed coupler. The design of this cell required 3-d simulations. MAFIA[4] and HFSS[5] were used for this purpose. The coupling ratio, β, was chosen to be 1.8 in order to shorten the RF filling time to 60 ns. The RF pulse width will be 150 ns or 2.5 filling times.

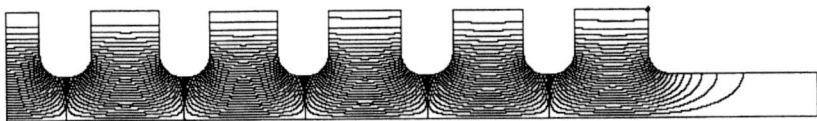

FIGURE 4. SUPERFISH simulation of RF gun

Many PARMELA[6] simulations were used to fine-tune the dimensions of the RF gun as well as optimize the field strength of the Solenoid for minimal transverse emittance It was also used to optimize the location of the accelerator and determine the relative field strengths and locations of the quadrupole triplet for a minimal spot size at the Interaction Region. An early example of these simulations is shown in figure 5. which shows the behavior of the beam profile along the beam axis.

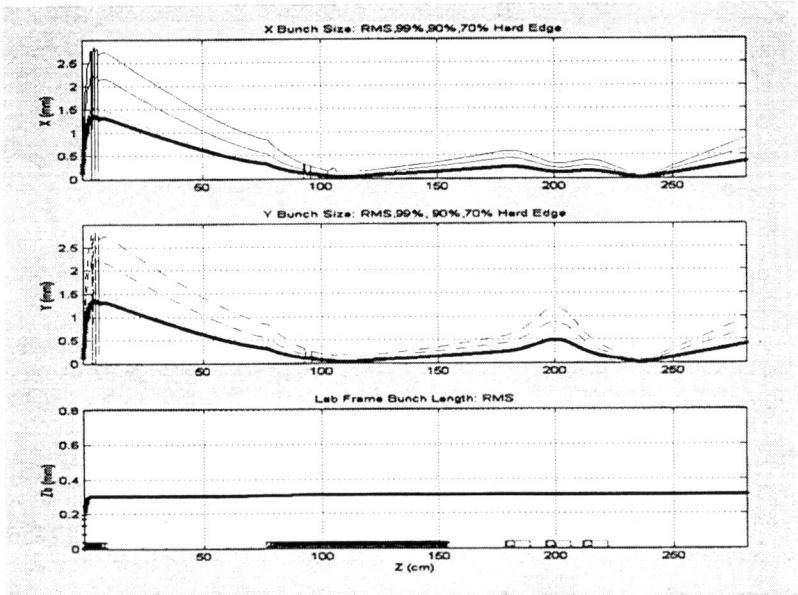

FIGURE 5. PARMELA simulation of RF gun

In this simulation the emittance at the interaction point was ~1.4 π-mm-mR with an RMS radius of ~14 microns. The shape of the pulse was nearly ideal (beer-barrel) with duration of .5ps and a cutoff radius of .25 mm (defined below). Because the laser energy density at these dimensions exceeded the damage threshold for copper [7], both the pulse width and duration were doubled in later studies. This change had the beneficial results that the gun could be operated well below the damage threshold and the quality of the beam could be improved. Further simulations, using more realistic (Gaussian) pulse shapes, resulted in good emittance beams at the Interaction Region. Table 2 shows a comparison of emittance for a beam with an initially "flat" shape with that of a Gaussian shape. [Note that in PARMELA, the initial beam shape, both temporally and transversely, is defined in terms of a Gaussian rms width, σ, and a cutoff width, σ_c. A flat beam is obtained by making $\sigma >> \sigma_c$. (Even for $\sigma = 2*\sigma_c$ the pulse shape is nearly flat).]

The effect of magnet transverse offset was also investigated. Using an option in our version of PARMELA we were able to displace the magnet transversely and observe the displacement of the beam centroid as well as the emittance change. The

TABLE 2. Comparison of beam characteristic at interaction region for different initial beam profiles.

	"Flat" Beam	Realistic beam
σ(radial)	1.0mm	0.25mm
σ_c(radial)	0.5mm	0.5mm
σ(temporal)	0.8ps	0.2ps
σ_c(temporal)	0.4ps	0.4ps
Emittance (rms)	0.72 π-mm-mR	1.1 π-mm-mR
Beam radius at Interaction Region	11.6 μ	22.6μ

values were determined at the waist of the beam. The results are shown in figure 6. It is interesting to note that for a magnet displacement $\leq$ 0.005 inches there is considerable beam centroid displacement (600 microns) but very little emittance degradation This would indicate that corrector coils could be used to bring the beam back on-axis as it enters the accelerator.

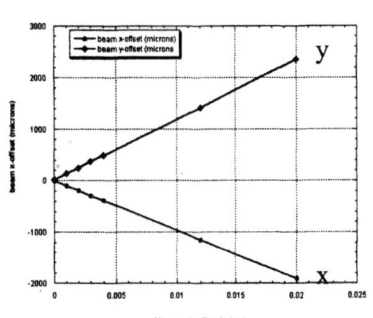 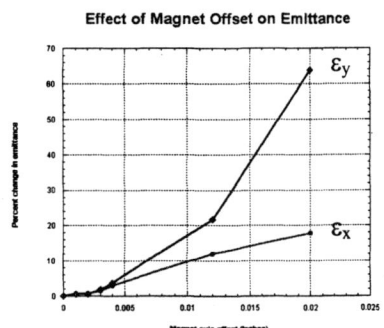

FIGURE 6. Effect of magnet offset on beam steering and emittance

COLD-TEST PROTOTYPE

A complete set of copper parts was fabricated using the dimensions obtained from SUPERFISH and MAFIA. Also fabricated were a pair of end caps, which permitted frequency measurements of individual cells. Half-cell end caps were first tuned so that the π-mode resonant frequency was 11.424 GHz and then used as matching cells for the other cells. The procedure is shown in figure 7.

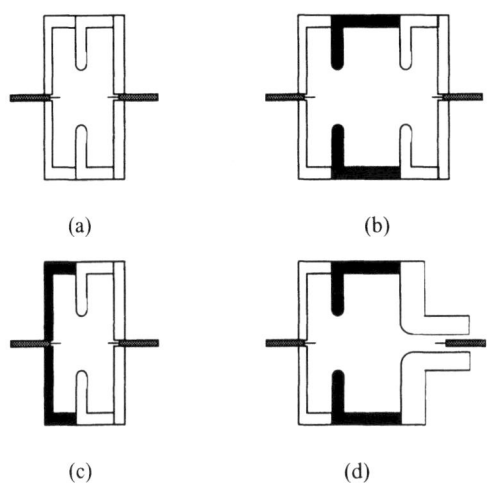

FIGURE 7. Sequence of cavity tuning. (a) Special end caps are tuned. (b) cells 2-4 are tuned individually. (c) Cell 1 (half-cell) is tuned. (d) Coupler cell is tuned. Coupling irises are not shown. The RF is coupled into/out of the cells by coaxial cables

After completion of this tuning, the cells were assembled as a unit and remeasured. The coupler cell was then tuned to correct the external Q. The resultant resonances and field profile is shown in figure 8. The field profile is obtained using the "bead-pull" method through a small hole in the center of the cathode.

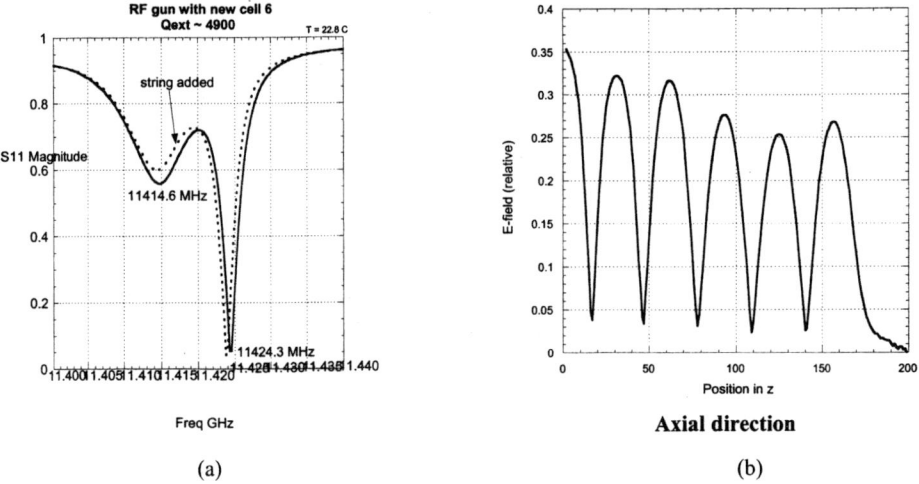

FIGURE 8. Structure before bonding. (a) Frequency spectrum. (b) Spatial field profile

In order to find out whether the cavity properties would change with bonding, the cells were high-temperature bonded and waveguides were brazed into the gun. The final cold test measurements are shown in figure 9.

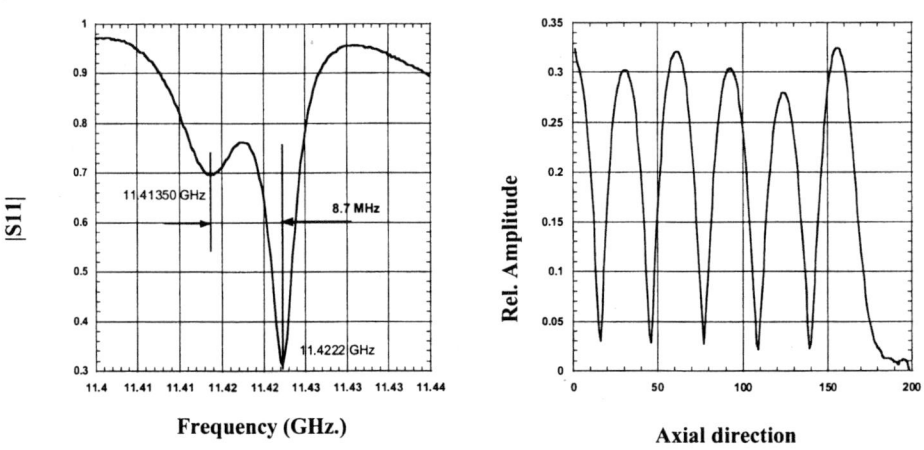

FIGURE 9. Structure after bonding

As can be seen from figure 9, the main change to the structure after bonding/brazing was an overall frequency shift of 2 MHz. This effect has been taken into consideration in the final gun dimensions.

SUMMARY

A 5.5 cell X-band RF Gun is being developed at SLAC as part of the development of a compact monochromatic X-ray source. Simulations using SUPERFISH and POISSON have been used to design the RF properties of the cavity cells and Solenoid magnet. MAFIA and HFSS were used where 3-d effects were present. PARMELA has been used to simulate and optimize the beam properties of the complete system. A cold-test prototype has been built and has been tuned to operate at the correct frequency and with a flat RF profile. The resultant dimensions will be used in constructing a working rf gun in the near future.

ACKNOWLEDGMENTS

One author,(aev) would like to thank Dr. E. Colby for his helpful suggestions and support throughout development of this work. We would also like to thank Dr. Cho Ng for his assistance with MAFIA simulations. This work supported by Department of Energy contract DE-AC03-76SF00515

REFERENCES

1. Solberg, T.D., A. Norman (2001). "Monoenergetic x-rays improve the therapeutic ratio in x-ray phototherapy of brain tumors." *Proceedings of the Workshop on the Medical Applications of Synchrotron Radiation*, Grenoble, France, March 1 – 3 2001
2. E. Burattini, "Synchrotron Radiation: A new trend in X-Ray Mammography", *ACTA Physica Polonica*, **A91**(4), 707-713, (1997)
3. James Billen and Lloyd M. Young, *POISSON, SUPERFISH reference manual*, LA-UR-96-1834.
4. R.Klatt, F.Krawczyk, W.R.Novender, C.Palm, T.Weiland, "MAFIA- A three-Dimensional Electromagnetic CAD ystem for Magnet, RF Structures and Transient Wake-Field Calculations", *Reports at the 1986 Linac Accelerator conference*, Stanford, USA June 2-6,1986
5. Agilent High-Frequency Structure Simulator v5.6, *reference manual* 85180-90194
6. Lloyd Young, PARMELA *reference manual*, LA-UR96-1835,January 8,2000,. The version of PARMELA used in this work is a modified version of PARMELA due to E. Colby, at SLAC.
7. T. Srinivasan-Rao, J. Fischer, and T. Tsang, "Photoemission studies on metals using picosecond ultraviolet laser pulses", *J. Appl .Phys.* **69** (5) 1 March 1991, 3291-3296.

Design of High-Power, MM-Wave Traveling-Wave Tubes

Bruce E. Carlsten, Lawrence M. Earley, W. Brian Haynes, and Robert M. Wheat

*Los Alamos National Laboratory
Los Alamos, NM 87545*

Abstract. Simulations have indicated that emerging electron sheet-beam technology can drive simple rippled and stepped waveguide traveling-wave tubes with extremely high gain. There are many design possibilities that need to be evaluated. The interaction can be made with the $n=-1$ (backward wave), $n=0$ (fundamental forward wave), or $n=+1$ (first space harmonic forward wave) interactions. In this paper, we discuss some of the fundamental design issues.

INTRODUCTION

An earlier analysis of planar, rippled waveguide traveling-wave tubes showed large gain for nominal structure and beam parameters [1]. In Figure 1 we show the geometry used for the rippled waveguide analysis. The beam travels along the axial direction, towards the right. The ripples are in the vertical dimensions and there is symmetry along the horizontal dimension. An infinite planar electron beam is assumed. The analysis starts with expanding the horizontal magnetic field in terms of the free-space modes, where $k_n = k + 2\pi n / \lambda$:

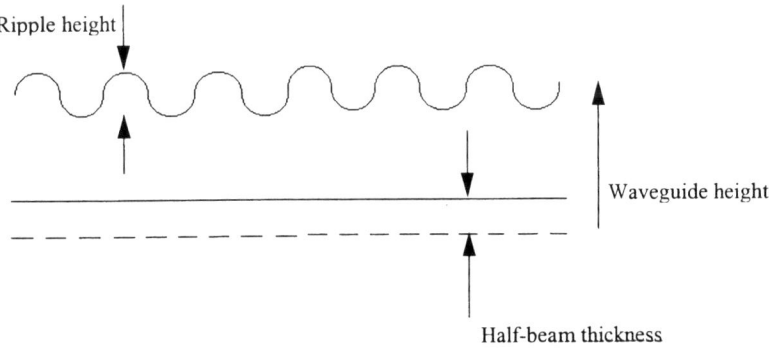

FIGURE 1. Notional geometry for a rippled waveguide traveling-wave tube

$$B_x = \sum_{n=-\infty}^{\infty} (a_n \cos(h_n y) + b_n \sin(h_n y)) e^{-jk_n} e^{j\omega t}. \tag{1}$$

We find a relationship between the modes by using Gauss' law to establish the boundary conditions at the beam and the boundary condition for the tangential electric field at the wall.

In Figure 2 we plot gain for the $n=1$ (forward wave) synchronism, for a nominal 140 kV, 15 A sheet beam with a 2 by 8 mm transverse size [2]. This example brings forth many of the interesting features of this class of traveling-wave tubes. First note that the gain at 300 GHz is about 1 dB/mm, or on the order of a dB/period (which is consistent with coupled-cavity traveling-wave tubes). Note that the period, and thus feature size, is on the order of 0.25-0.5 mm. Emerging lithography technologies based on third generation light sources, such as LIGA [3], are capable of producing high-quality planar structures with dimensional accuracies on the order of 1 µm, and can produce slow-wave structures with these feature sizes.

The following sections in this paper will be focused on various design issues of these types of traveling-wave tubes. In particular, the analysis will be focused on a 100 GHz experiment at Los Alamos intended as a technology step to the ultimate goal of 300 GHz. The next section will provide a discussion of the dispersion relation of a ridged waveguide and of possible interaction modes (backwards and forwards), including space-harmonic amplitudes. The section after that will describe gain calculations using a modal analysis of the ridged waveguide.

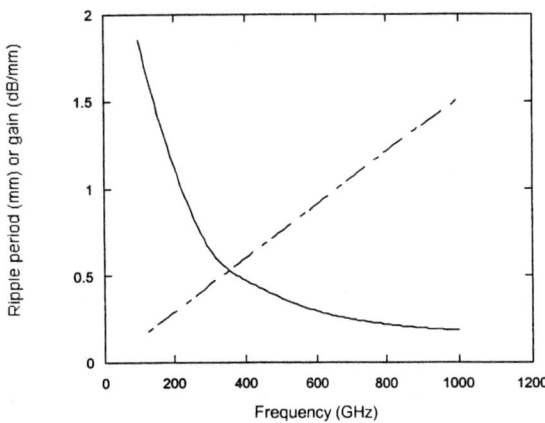

FIGURE 2. Solid line is ripple period, dashed line is gain - for 2 x 8 mm geometry, 140 kV, 15 A sheet beam, 5% ripple amplitude

DISCUSSION OF INTERACTION MODES

The main design goal is to make the electron beam synchronous with a large-amplitude space harmonic. From Eqn. (1) we see that the nth space harmonic will have phase velocity $\omega/(k + 2\pi n/\lambda)$, where k is the fundamental mode propagation constant. For a smooth waveguide, ω/k is always greater than the speed of light, c. The phase velocity of a space harmonic can be made synchronous with an electron beam by either going to a nonzero n (typically either the +1 or –1 modes), or by deforming the waveguide such that ω/k itself is less than c.

In particular, we will consider a 100 GHz example because of a planned experiment at Los Alamos – a backward interaction at 100 GHz with a 10 kV electron beam. We will start by first considering the waveguide dispersion relation, with and without ridges.

Waveguide Dispersion Diagrams

Let us consider a planar waveguide with a vertical 2 mm gap (and infinite in the transverse x direction). We assume the waveguide fields have a $e^{j(\omega t - kz)}$ dependency, using standard notation. The dispersion relation defines k in terms of ω (or vice versa), and this relationship is plotted in the dispersion diagram. For a simple planar waveguide of half height y_0 this is easy – the dispersion relation is given by

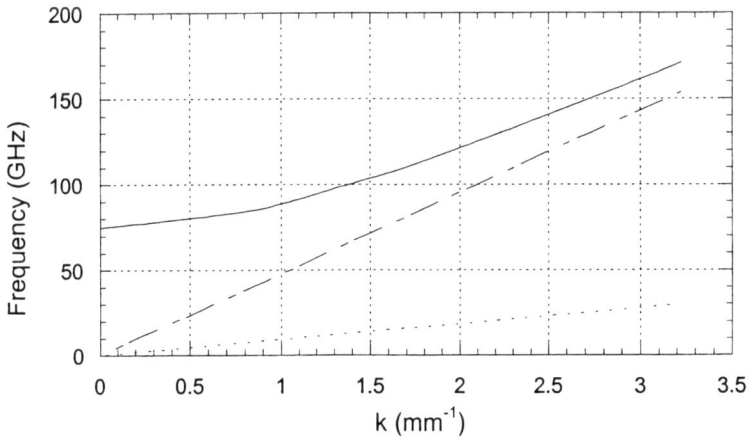

FIGURE 3. Smooth waveguide dispersion curve, compared to the speed of light and a 10 kV beam.

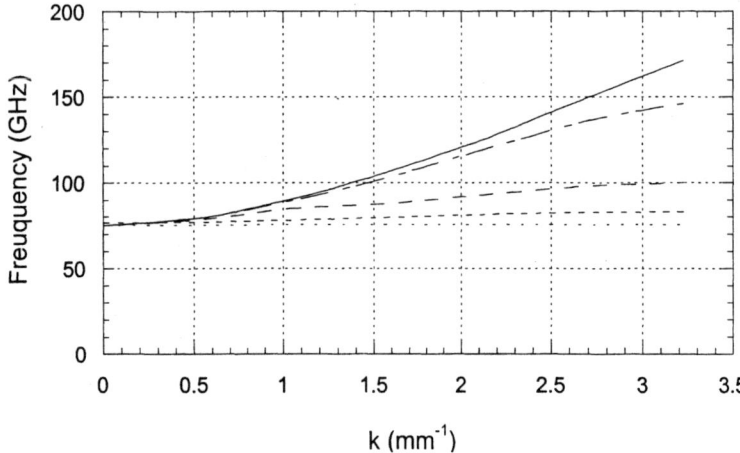

FIGURE 4. Waveguide dispersion curve for vanes with a 1 mm period, varying in height from 0.2 mm to 0.9 mm.

$$k^2 = \frac{\omega^2}{c^2} - \left(\frac{\pi}{2y_0}\right)^2 \qquad (2)$$

(this equation comes directly from the wave equation using the lowest order transverse mode). We plot this dispersion relation, for a 2 mm total gap, in Figure 3. The solid line corresponds to the relation between k and ω. Noting that ω/k is in fact the phase velocity, the dot-dashed line corresponds to a phase velocity of the speed of light c and the dotted line corresponds to a phase velocity equal to 5.85 (10^7) m/sec (the speed of a 10 kV electron beam). In order to have synchronism with a 10 kV beam at 100 GHz, the dotted line needs to intersect the waveguide mode at that frequency.

We can deform the waveguide mode by putting in periodic ridges (which are called vanes if they are relatively very narrow). With vanes about 1 mm apart, we find the curves shown in Figure 4. The solid line is for the smooth case, the dot-dash line has 0.2 mm steps, the long dashed line has 0.5 mm steps, the small dashed line has 0.7 mm steps, and the dotted line has 0.9 mm steps. The dispersion relation is symmetric about a k value of π/λ, which here is k=3142 m^{-1}.

We show the dispersion relation between a k of zero and $3\pi/\lambda$ in Figure 5 (this is for a period λ of 0.73 mm). The region between $(0, \pi/\lambda)$ is known at the fundamental mode, the region between $(\pi/\lambda, 2\pi/\lambda)$ is known at the n=-1 backward mode, and the region between $(2\pi/\lambda, 3\pi/\lambda)$ is known at the n=+1 forward mode. The apparent inconsistency with the backward wave is resolved by recognizing it is the n=-1 space harmonic of the fundamental component lying in the interval $(-\pi/\lambda, 0)$.

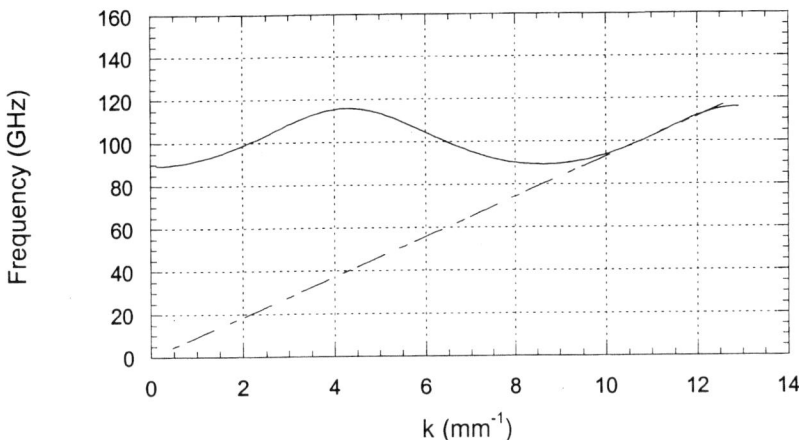

FIGURE 5. Waveguide dispersion relation showing synchronism between a 10 kV electron beam and the $n=1$ space harmonic at 100 GHz.

An rf wave at 100 GHz will have components of all the space harmonic modes, and the rf field will be a linear superposition of these modes, with relative amplitudes determined by the boundary conditions.

N=0 Synchronism

In order to be synchronous with the fundamental forward mode, the period of the steps needs to be very short and the vane height needs to be large in order to flatten the smooth waveguide mode. We can find the period by imaging the waveguide is a line of accelerator cells and requiring that the phase shift between cells is:

$$p = v_b \tau \frac{\phi}{2\pi} , \qquad (3)$$

where p is the period, v_b is the beam velocity, τ is the rf period and ϕ is the phase shift. The required periods are shown in Table 1. These are small numbers, but are within the capability of LIGA manufacturing. However, for interaction at 100 GHz with a 10 kV electron beam, it is worth considering a higher-order space-harmonic interaction for a initial experiment.

TABLE 1. Periods required for fundamental mode synchronism

Phase shift:	100 GHz, 10 kV	100 GHz, 140 kV	300 GHz, 140 kV
$\pi/3$	0.0975 mm	0.310 mm	0.103 mm
$\pi/2$	0.146	0.464	0.155
$2\pi/3$	0.195	0.619	0.206

N=1 Synchronism

In addition to alleviating fabrication issues, going to an $n=1$ space-harmonic interaction can also greatly increase bandwidth by matching the slopes of the rf dispersion curve and the beam curve at the interaction point. We can demonstrate this by considering the dispersion relation for a theoretical chain of coupled cavities, which can be represented by [4]

$$\omega = \frac{\omega_0}{\sqrt{1+\kappa \cos(kp)}} = \frac{\omega_c \sqrt{1+\kappa}}{\sqrt{1+\kappa \cos(kp)}}, \qquad (4)$$

where ω_0 is the center of the passband, κ is the coupling coefficient (determined by the size of the steps), and ω_c is the lower cutoff frequency. This has maximum positive slope of $\omega_0 \kappa p/2$ at position $(\omega,k)=(\omega_0, \pi/2p+2\pi n/p)$, for $n=0,1,2$, and so on. We find maximum bandwidth if the slope is matched at an intersection, or if $\omega_0 \kappa p / 2 = \omega_0 / (5\pi/2p) = v_b$. Synchronism leads to a period p of 0.73 mm and matching slopes gives a coupling factor of $\kappa = 4/5\pi$. This matched condition was shown in Figure 5.

We can verify that this coupling coeffient is reasonable by considering the maximum coupling coefficient that corresponds (approximately) to the case with very small steps. For that case, the top frequency of the passband is given by $\omega_c^2 + \left(\frac{\pi}{p}\right)^2$.
The corresponding maximum coupling coefficient must then be given by

$$\omega_0^2 \left(\frac{1}{1-\kappa_m} + \frac{1}{1+\kappa_m}\right) = \left(\frac{\pi}{p}\right)^2 \qquad (5)$$

which has solution

$$\kappa_m = \frac{(2y_0/p)^2}{2+(2y_0/p)^2}. \qquad (6)$$

For the $n=1$ example we are considering, y_0 is about 1 mm and p is about 0.73 mm, so the maximum coupling coefficient is about ½, larger than required for the matched solution of $\kappa = 4/5\pi$.

With reasonable gains, we can expect 10%-20% bandwidth. We can verify this effect by calculating the dispersion curve using the code SUPERFISH [5]. With a real geometry, we pick up some nonideal effects, but we basically get the same result. The SUPERFISH case has period of 0.73 mm, gap half height of 0.87 mm (the ideal

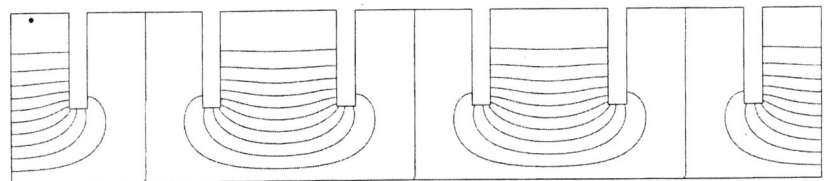

FIGURE 6. SUPERFISH simulation of nominal 100 GHz geometry, showing $\pi/2$ mode.

using $\omega_c = c\pi/2y_0$ would be 0.839) and step height of 0.38 mm (to get about the right coupling). We can verify the $\pi/2$ mode is at 100.4 GHz by looking at the field plot, shown in Figure 6. The SUPERFISH dispersion diagram with beam curve is shown in Figure 7. This curve shows much of the features of the theoretical curve, but with the addition of nonideal coupling. The coupling coefficient can be increased more for better bandwidth (at the expense of decreasing the beam pipe slot with more potential beam interception and less gain).

Comparison of Mode Coefficient Amplitudes

We have written the code DETER, based on a modal analysis of the ridged waveguide geometry, to solve the disperson relation with and without beam [6]. In Table 2 we show the space-harmonic mode amplitudes found with DETER, for the case shown in Figure 7.

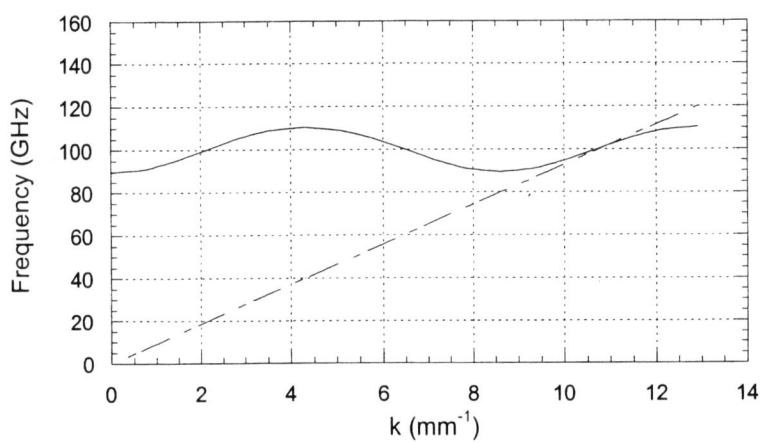

FIGURE 7. SUPERFISH dispersion curve showing wide bandwidth $n=1$ interation with 10 kV beam.

TABLE 2. Mode Amplitudes for $n=1$ interaction.

Mode	Complex amplitude
-4	(9.21e-6, 6.07e-4)
-3	(-7.17e-3, 2.50e-2)
-2	(-0.00426, 0.004955)
-1	(-0.129, 0.0593)
0	1
1	(0.0233, 0.0107)
2	(4.86e-2, 5.66e-2)
3	(2.49e-4, 8.66e-4)
4	(4.38e-6, -2.88e-5)

The negative space harmonics have both negative group and phase velocities – however the $n=-1$ mode is equivalent to the first backward wave (negative group velocity) with positive phase velocity. We note that its amplitude it much larger than any of the forward modes, except the $n=0$ mode.

We can estimate the ratio of the amplitudes of the different space harmonics in the limit that there is very little coupling between cells and they are essentially separate pill box cavities. Consider the case where the mth cell starts at $z_m = m\lambda$. Now the axial electric field in the cell can be described in terms of the space harmonics as

$$E = \sum_{n=-\infty}^{\infty} E_n e^{-j\left(k + \frac{2\pi n}{\lambda}\right)z} e^{j\omega t} \qquad (7)$$

and in terms of a pillbox cavity as

$$E = E_0 \, \text{Re}\{e^{j\omega t + \phi}\} \qquad (8)$$

where we will use the arbitrary phase notation that $\phi = mk\lambda$. Multiplying both equations by $e^{j\left(k + \frac{2\pi n'}{\lambda}\right)z} e^{-j\omega t}$ and integrating over a cell length λ we immediately find

$$E_n = \frac{jE_0}{\phi + 2\pi n}\left(1 - e^{-j\phi}\right). \qquad (9)$$

The largest space harmonic is always the $n=0$ one, but the $n=-1$ one is always second largest. It can in fact equal the amplitude of the fundamental mode as the phase shift per cell approaches π.

TABLE 3. Relative space harmonic mode amplitudes

	$\pi/2$	π
$\|E_{-2}/E_0\|$	1/7	1/3
$\|E_{-1}/E_0\|$	1/3	1
$\|E_0/E_0\|$	1	1
$\|E_1/E_0\|$	1/5	1/3
$\|E_2/E_0\|$	1/9	1/5

In Table 3 we summarize the space harmonic mode amplitudes for phase shifts of both $\pi/2$ and π.

GAIN PREDICTIONS

After reviewing the relative mode amplitudes, we have decided to focus on the $n=-1$ backward wave interaction for our demonstration experiment because the deterioration in gain for the $n=1$ interaction was too severe for any increase in bandwidth. Using DETER, we can estimate the gain of the $n=-1$ interaction using a 0.5 A, 10 kV electron beam in a 0.5-cm wide structure, which is shown in Figure 8. For this example, we used a vane period of 0.508 mm, a vane separation of 0.381 mm, a vane height of 0.11 mm, and a centerline-to-vane-tip spacing of 1.033 mm. Note the extremely high gain over the entire bandwidth of the slow-wave structure. In Figure 9, the 100 GHz structure is compared to a penny.

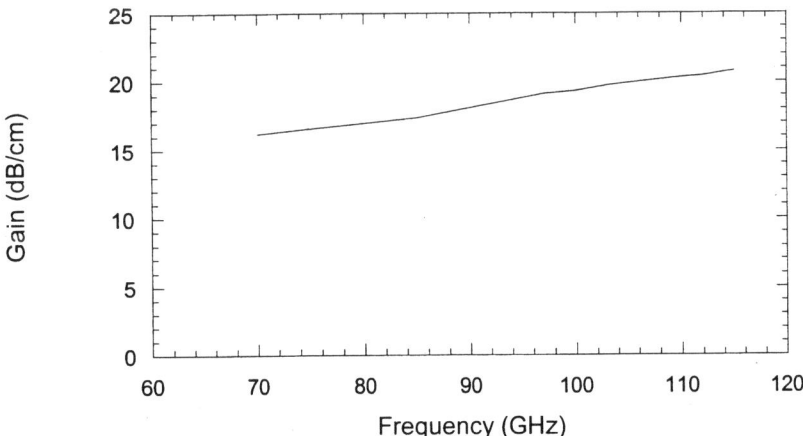

FIGURE 8. Gain for $n=-1$ interaction for nominal 100 GHz geometry and 0.5 A, 10 keV electron beam.

FIGURE 9. Nominal 100 GHz structure for technology experiment. The entire structure is 0.24" by 0.24" by 3" long, with ½ mm period. Input and output waveguides will be run inside the top and bottom structures.

SUMMARY

We have considered various design possibilities for rippled waveguide and ridged waveguide traveling-wave tube amplifiers, and in particular we have tried to contrast the forward n=0 and n=1 modes and the backward n=-1 mode.

We anticipate the highest gain for the n=0 case, which leads to very short structure periods. For high beam energy at 300 GHz, using advanced lithography fabrication techniques, such structures can be fabricated. However, for lower electron beam energy (say 10 kV), either the $n=1$ or $n=-1$ interactions need to be used. We have shown that the interaction strength is much lower for the $n=1$ interaction than for the $n=-1$ interaction. The $n=-1$ interaction has very high gain, and very reasonable bandwidth, and is a good candidate for an exploratory experiment leading to a final goal of a 300-GHz traveling-wave tube.

ACKNOWLEDGMENT

This work was supported by the Los Alamos National Laboratory Directed Research and Development program, under the auspices of the U.S. Department of Energy.

REFERENCES

1. Carlsten, B. E., *Phys. Plasmas,* **8**, 2702 (2001).
2. Colby, E. R., Caryotakis, G., Fowkes, W. R., and Smithe, D. N., "W-Band Sheet Beam Klystron Simulation" in *High Energy Density Microwaves*, edited by R. Phillips, AIP Conference Proceedings 474, New York, 1998, 74.
3. J. Hruby, "Overview of LIGA Microfabrication," these proceedings.
4. Wangler, T., *RF Linear Accelerators*, John Wiley and Sons, Inc., New York, 1998, chapter 3.
5. Young, L., private communication (2002).
6. Carlsten, B. E., "Modal Analysis and Gain Calculations for a Sheet Electron Beam in a Ridged Waveguide Slow-Wave Structure," submitted to *IEEE Trans. Elec. Dev.*

Theory and Experiment of Ultra High Gain Gyrotron Traveling-Wave Amplifier

K. R. Chu*, T. H. Chang*, L. R. Barnett*, and S. H. Chen[†]

*Department of Physics, National Tsing Hua University
Hsinchu 300, Taiwan
[†]National Center for High-Performance Computing
Hsinchu 300, Taiwan

Abstract. The gyrotron traveling-wave tube (gyro-TWT) is a millimeter-wave amplifier based on the electron cyclotron maser (ECM) instability. High power and broad bandwidth capability makes it an attractive source of coherent radiation. The current paper presents a brief overview of the gyro-TWT research over the past quarter of a century. Advances made by different groups employing various schemes are discussed and achieved performances are surveyed. This a followed by a focused study of an ultra high gain scheme as an illustration of the various stability issues involved.

INTRODUCTION

Traveling-wave amplification via the ECM instability was demonstrated in the mid-1960s in a crossed-field device (trochotron) employing an electron beam in which the electrons move along the waveguide in E×B drift motion [1]. However, the gyro-TWT in its present form [2] employs an electron beam comprised of helically moving electrons along a static magnetic field. It evolved from a series of experiments [3-8] aimed at high power microwave generation by intense relativistic electron beams. In the gyro-TWT as in other gyro-devices, the free energy resides in motion transverse to the applied magnetic field. An injected wave of low power level is amplified through the ECM instability along the length of a waveguide (Fig. 1).

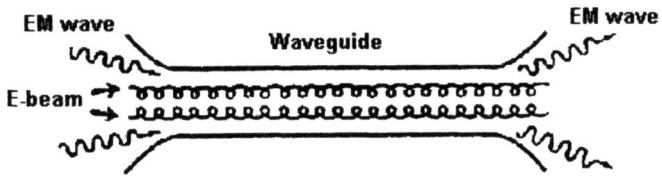

FIGURE 1. Schematic of the gyrotron traveling-wave amplifier.

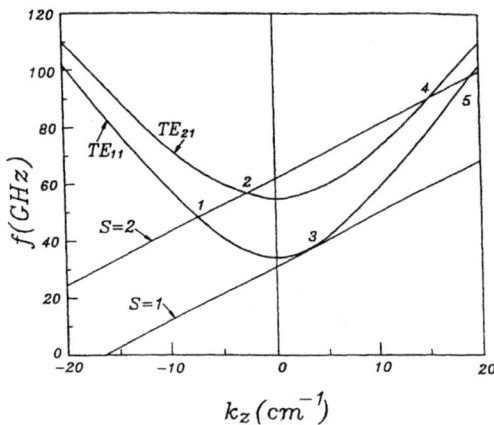

FIGURE 2. $\omega-k_z$ diagram of a fundamental harmonic gyro-TWT based on the TE_{11} mode convective instability (point 3). Absolute instabilities on backward waves (points 1 and 2) and other convective instabilities (points 4 and 5) are also indicated.

In the gyro-TWT, the electrons are in resonance with a multitude of forward- and backward-waves both at the fundamental and higher cyclotron harmonics. This is illustrated in Fig. 2 which plots the ω-k_z diagram of the TE_{11} and TE_{21} waveguide modes (for a waveguide radius of 0.27 cm) and the fundamental (s=1) and second (s=2) cyclotron harmonic beam-wave resonance lines. Intersections in the forward-wave region (points 3, 4, and 5) are convective instabilities, which grow and propagate along with the electron beam. Operating condition of the gyro-TWT is normal at a grazing intersection point (point 3). Interactions in the backward wave region (points 1 and 2) are absolute instabilities [9-11], which grow locally from the noise level by way of an internal feedback loop. Hence, the system can easily break into oscillations. The problem of backward-wave oscillation is compounded by an additional source of oscillation that results from end reflections. These oscillations will either interfere with the amplification process or prevent the gyro-TWT from operating in the optimum parameter regime. Understanding of various types of oscillations and methods for their suppression constitute the principal challenges of the gyro-TWT research.

SURVEY OF GYRO-TWT RESEARCH

Early ECM experiments employing an intense relativistic electron beam was found to saturate at relatively low efficiency, both in theory [12] and in experiment [8]. However, the experiment in [8] provided a strong impetus for the development of practical traveling-wave amplifiers based on the ECM interaction. The pursuit of the gyro-TWT gained additional momentum though theoretical studies of the saturation mechanisms and efficiency scaling which predicted an energy conversion efficiency as high as 60% (in the beam frame) for moderately energetic electron beams [13].

Further studies in the cylindrical configuration led to the theoretical design [14,15] and first operation of the gyro-TWT at the Naval Research Laboratory [16-17]. In the limited space of this article, we will summarize the key parameters and achieved performances of this and subsequent experiments in Table I. As remarked in the last column, each experiment represents a significant step toward the realization of the ultimate potential of the gyro-TWT. For comparison, performance characteristics of a state-of-the-art Ka-band TWT are also listed in the first row of Table I.

Not surprisingly, the gyro-TWT was found to be susceptible to oscillations, which had severely restricted its performance. In a later experiment at NRL, stable saturated gain of over 40 dB was achieved in a distributed-loss circuit with improved bandwidth but significantly lower output power (Table I) [18] A broadband scheme [19-20] employing a linearly tapered waveguide with the magnetic field profiled in proportion to the cutoff frequency was also tested in the same apparatus (Table I) [21].

Gyro-TWT experiments were concurrently conducted at Varian Associates in the C-band [22-26] and W-band [26]. In the C-band experiments, impressive performances were achieved (Table I) with the application of 6-10 dB distributed loss over the first two thirds of the interaction waveguide to overcome oscillations. AM and PM modulation coefficients, spectral purity, phase linearity, and output noise have also been characterized for the C-band tube [25]. Reviews of the gyro-TWT experiments during this period can be found in [27] and [28].

Experimental (and to a large extent also theoretical) gyro-TWT activities subsided for a number of years following the early ground breaking experiments at NRL and Varian. A new round of international research activities started in the late 1980s, all with emphasis on the fundamental issues of stability and strongly supported by theories. The group at the National Tsing Hua University (NTHU) in Taiwan performed a series of TE_{11}-mode, Ka-band experiments for the study of mode competition in the gyro-TWT (Table I) [29-35]. These investigations led to a high gain scheme which will be discussed in the following section.

At the same time, the UCLA/UC Davis team pursued the physics and technology of harmonic interactions. Finite Larmor radii allow the electrons to sense the transverse variations of the rf fields. Hence, harmonic cyclotron interactions become possible and in some cases cause serious mode competition problems. However, they are also useful for the alleviation of the magnetic field requirement as well as generating increased peak power beyond that obtainable at the fundamental cyclotron interaction. The UCLA/UC Davis group reported the first operations of harmonic gyro-TWTs (Table I) [36-39] by employing an axis encircling electron beam generated by gyroresonant rf acceleration in a cavity [40]. By producing the highest gyro-TWT output power to date (207 kW at 16.7 GHz) for a moderately energetic electron beam (Table I), these experiments verified the prediction that the second harmonic gyro-TWT can be more stable to oscillations and generate higher powers than at the fundamental harmonic [41-43].

TABLE 1. Survey of experimental gyro-TWTs and comparison with a state-of-the-art commercial TWT

Institution [references]	Cyclotron harmonic no / mode	V_b (kV)	Center frequency (GHz)	Peak power (kW)	Saturated gain (dB)	Saturated efficiency (%)	Saturated bandwidth 3 dB (%)	Remarks
CPI [VTA 5701]	---	50	35	50	40	16	6	State-of-the-art Ka-band TWT
NRL [16-17]	$1/TE_{01}$	70	35	16.6	20	7.8	1.5	First demonstration of gyro-TWT
NRL [18]	$1/TE_{01}$	70	35	3.2	42	1.5	2	Gain enhancement with distributed wall losses
NRL [21]	$1/TE_{01}$	70	35	---	18 (linear)	---	13 (linear)	Bandwidth broadening with tapered waveguide and magnetic field
Varian [22-25]	$1/TE_{11}$	60	5	120	18	26	6	Distributed wall losses, AM/PM sensitivity, noise characterization
Varian [26]	$1/TE_{11}$	50	94	20	30	8	2	First attempt at the W-band (1982, unpublished)
NTHU, Taiwan [29-31]	$1/TE_{11}$	80	35	18.4	18	18.6	10	Systematic characterization of mode competition
NTHU, Taiwan [32]	$1/TE_{11}$	90	35	27	35	16	7.5	Improved stability with a sever
NTHU, Taiwan [33]	$1/TE_{11}$	100	35	62	33	21	12	Study of oscillation suppression with distributed wall losses
NTHU, Taiwan [34,35]	$1/TE_{11}$	100	35	93	70	26.5	8.6	Demonstration of an ultra high gain scheme employing distributed wall losses
UCLA [36]	$8/TE_{81}$	350	16.2	0.5	10 (linear)	1.35	4.3 (linear)	First proof-of-principle experiment on harmonic gyro-TWT
UCLA and UC Davis [38,39]	$2/TE_{21}$	80	15.7	207	16	13	2.1	Demonstration of stability and high power with harmonic interaction
NRL [44,45]	$1/TE_{10}$ (rectangular)	33	34	---	20 (linear)	---	33 (linear)	Record bandwidth, achieved with a single-stage tapered circuit
NRL [48]	$1/TE_{10}$	33	35	8	25	16	20	Broadband two-stage tapered circuit
NRL [49]	$1/TE_{11}$	900	35	20000	30	11	---	IREB-driven gyro-TWT
IAP, Russia and U Strathclyde, UK [52,53]	$2/TE_{11}+TE_{21}$	185	9.4	1100	37	29	21	Demonstration of broadband and highly efficient corrugated circuit

132

The NRL group, on the other hand, launched a research effort at the tapered gyro-TWT [44-48]. Record linear and saturated bandwidths were produced in single-stage and two-stage schemes, respectively (Table I). A separate NRL effort was directed at intense-relativistic-electron-beam (IREB) driven gyro-TWTs, employing a 900 kV electron beam [49,50]. It produced 20 MW output power in the Ka-band at a gain of 30 dB (Table I) [49] and 11% efficiency. This experiment has demonstrated the viability of the gyro-TWT as an ultra-high-power, broad-bandwidth, and phase-controlled source.

More recently, a Russian and United Kingdom team developed a helically corrugated interaction structure [51]. This novel circuit provides the important advantages of broadband operation and high tolerance to the electron velocity spread. Preliminary results of a second harmonic experiment confirmed theoretical predictions and demonstrated the feasibility of this promising scheme (Table I) [52,53].

A SCHEME FOR ULTRA HIGH GAIN

The distributed-loss structure employed previously to suppress oscillations [18, 22-25] can also yield ultra high stable gain. The scheme is based on the different responses to wall losses between the cold tube and hot tube modes. The cold tube mode has all of its energy in the electromagnetic fields. In a hot tube, however, energy of the beam-generated mode resides not only in the electromagnetic fields but also in the kinetic energy of the oscillatory motion of the electrons, the latter being an integral part of the hot tube mode. The lossy wall absorbs the electromagnetic energy, but not the oscillatory energy of the electrons. Thus, wall losses of the amplifier circuit attenuate the reflected wave (basically a cold tube mode) significantly more than they reduce the gain of the amplifying wave (a hot tube mode). It can be shown analytically [54] that the reduction in hot tube gain due to wall losses is only one third of the cold tube attenuation over the same distance.

Such unequal effects have been exploited to simultaneously achieve both high gain and reflective stability. In this scheme (Fig. 3), the lossy and conducting-wall sections comprise the linear and nonlinear stages of the amplification, respectively. The linear section is made sufficiently long to provide the desired gain, while the nonlinear section length is constrained to a minimum to enhance the threshold of absolute instabilities.

We first discuss a recent investigation of three types of oscillations by employing the trajectory tracing simulation technique [35]. Imposition of physical boundary conditions at both ends allows the evaluation of a self-consistent rf field profile, $f(z)$, to account for wave reflections at all interfaces and nonuniformities.

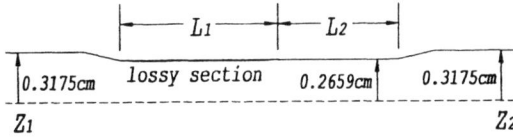

FIGURE 3. Schematic of a Ka-band, TE_{11} mode, fundamental harmonic gyro-TWT. Sections L_1 and L_2 form the interaction region. The ends are tapered for broadband coupling.

Reflective oscillations of a global nature start when the total gain exceeds the reflection at the input/output ends plus the attenuation in the lossy section (all in dB). Figure 4a shows the field profile of a typical global reflective oscillation. Calculations indicate that the oscillation can be stabilized by lowering the operating current (I_b) and magnetic field (B_o), or by increasing the wall resistivity. Beam current is usually fixed by the power requirement and the magnetic field must be fine tuned for maximum efficiency; hence, wall resistivity provides the most effective means for stabilization. Wall resistivity reduces the gain, but this can be easily compensated by a lengthened lossy section.

The conducting-wall section (of length L_2 in Fig. 2) by itself is subject to localized oscillations due to reflections at the lossy-wall junction on the left and the output structure on the right. Figure 4b illustrates the field profile of such an oscillation in the lowest order (TE_{111}) axial mode. Since the oscillation is localized to the conducting-wall section, the oscillation power is found to be nearly independent of the length (L_1) and wall resistivity (ρ) of the lossy section. These two features are in contrast to the high sensitivity of the global reflective oscillation to L_1 and ρ.

The gyro-TWT shown in Fig. 2 is most susceptible to the TE_{21} mode absolute instability at the second cyclotron harmonic [29-31]. Figure 4c illustrates the field profile of such an oscillation. Again, the wall resistivity is found to be highly effective in stabilizing the absolute instability.

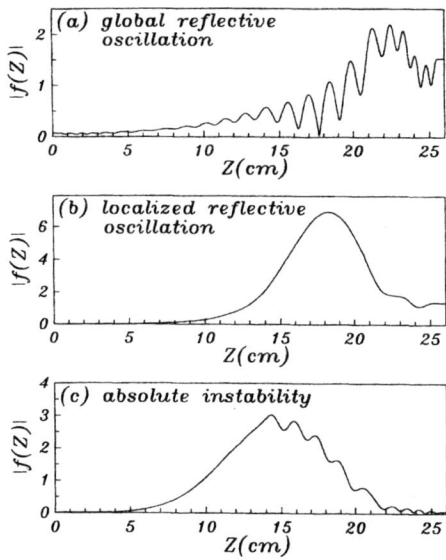

FIGURE 4. Calculated profiles of the RF field amplitude $|f(z)|$ in the structure of Fig. 3 for (a) the global reflective oscillation, (b) the localized reflective oscillation, and (c) the absolute instability.

Of the three types of oscillations, only the localized reflective oscillation cannot be effectively suppressed by distributed wall losses. It has indeed been observed in the experiment described below, but at a magnetic field value above the optimum range for the amplifier operation.

An experimental gyro-TWT was assembled to verify the ultra high gain scheme just described [34,35]. A mechanically tunable magnetron injection electron gun was attached to the interaction structure of Fig. 1. Lengths of the graphite-coated lossy section (L_1=20 cm with ~100 dB loss) and the conducting-wall section (L_2=4 cm) were chosen to achieve high gain as well as stability. Input/output waves were coupled at z_1 and z_2 through the side walls with newly designed oscillation-free couplers which also function as converters between circularly and linearly polarized waves. The magnetic field was provided by a superconducting magnet system. Output power at low duty was measured with a calibrated crystal detector (with estimated accuracy of ±5%) and verified with a calorimeter (agreement was within ~5%). At the operating beam current of 3.5 A, the gyro-TWT was found to be zero-drive stable from all three types of instabilities in the optimum range of the operating magnetic field (12.65 kG < B_o <12.75 kG). As the magnetic field was increased, a localized reflective oscillation was observed and identified to be the TE_{111} mode of the conducting-wall section.

Figure 5 plots the saturated output power and gain (dots) as functions of the frequency. The peak power of 93 kW corresponds to a saturated gain of 70 dB and efficiency of 26.5%. The ultra high gain, 30 dB beyond that previously achieved (Table 1), permits the use of solid state sources as drives. The full-width half-maximum bandwidth is 3 GHz, approximately 8.6% of the center frequency. Measured data are closely matched by theoretical predictions (solid line) using the simulated beam parameters: α=1, $\Delta v_z/v_z$ = 5%, and r_c = 0.09 cm, where $\Delta v_z / v_z$ is the electron velocity spread and r_c is the radial position of the electron guiding centers. The theory also predicts that the saturated power is almost independent of the length of the lossy section while the gain is linearly proportional to it.

The peak Ohmic power dissipated on the waveguide is calculated to be approximately 10 kW. With profiled wall losses, it can be evenly distributed over the lossy surface area with a peak dissipation rate of ~300 W/cm^2. The average-power handling capability will be limited by the availability of proper heat-resistant lossy materials and advanced cooling techniques. However, supplementary attenuation by broadband side-wall coupling to an external load could conceivably be implemented to remove this limitation.

Figure 6 shows the measured output power versus the input power (dots). Linear and saturated behaviors are consistent with the calculated data (solid line). Again, α=1, $\Delta v_z/v_z$ = 5%, and r_c=0.09 cm were assumed in the calculations. In all the measurements for Figs.5 and 6, we have not detected any spurious oscillation.

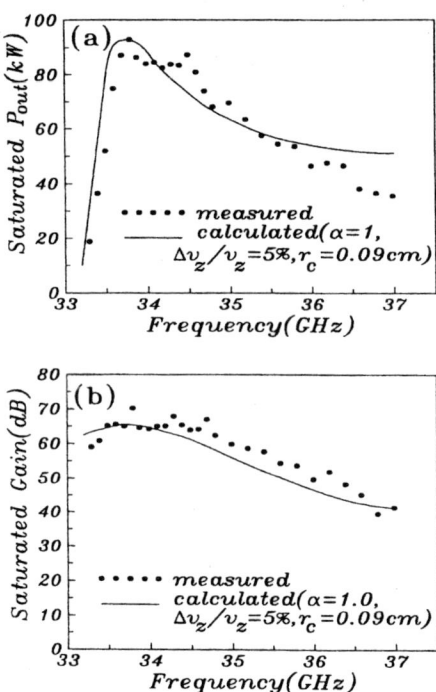

FIGURE 5. Saturated output power (a) and gain (b) versus the frequency. Measured and calculated data are shown by dots and lines, respectively. V_b =100 kV, I_b=3.5 A, and B_o=12.7 kG.

Figure 6 shows the measured output power versus the input power (dots). Linear and saturated behaviors are consistent with the calculated data (solid line). Again, $\alpha=1$, $\Delta v_z/v_z = 5\%$, and r_c=0.09 cm were assumed in the calculations. In all the measurements for Figs. 5 and 6, we have not detected any spurious oscillation.

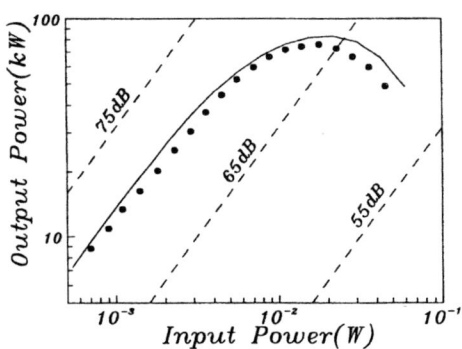

FIGURE 6. Measured (dots) and calculated (lines) output power versus the drive power. V_b =100 kV, I_b=3.5 A, B_o=12.7 kG, and f=34.2 GHz.

CONCLUSION

We have described the physics and technology issues of the gyro-TWT as well as highlights of the worldwide research in tackling these issues over a long period of time. Studies indicate that a basic understanding of the intricate interplay between the absolute/convective instabilities, circuit losses, and reflective feedback is of fundamental importance to the scientific demonstration of the potential capability of the gyro-TWT. A performance survey indicates that these efforts have culminated in the demonstrations of the gyro-TWT as a practical millimeter-wave radiation source of unprecedented power, gain, and bandwidth.

ACKNOWLEDGMENT

The author is grateful to Dr. L. R. Barnett, Dr. S. H. Gold, Prof. V. L. Granatstein, Prof. N. C. Luhmann, Jr., and Dr. D. B. McDermott for their critical comments. This work was sponsored by the National Science Council, Taiwan.

REFERENCES

1. Antakov, I. I., Gapanov, A. V., Malygin, O. V., and Flyagin, V. A., *Radio Engineering and Electron Physics* 11, 1195-1197 (1966).
2. Granatstein, V. L., Sprangle, P. S., Drobot, A. T., Chu, K. R., and Seftor,, J. L., *U.S. Patent*, no. 4224576 (1980).
3. Nation, J. A., *Appl. Phys. Letters* 17, 491-494 (1970).
4. Friedman, M., and Herndon, M., *Phys. Rev. Lett.* 28, 210-212 (1972).
5. Carmel, Y., and Nation, J. A., *J. Appl. Phys.* 44, 5268-5274 (1973).
6. Friedman, M., Hammer, D. A., Manheimer, W. M., and Sprangle, P., *Phys. Rev. Lett.* 31, 752-755 (1973).
7. Granatstein, V. L., Sprangle, P., Parker, R. K., and Herndon, M., *J. Appl. Phys.* 46, 2021-2028 (1975).
8. Granatstein, V. L., Sprangle, P., Herndon, M., Parker, R. K., and Schlesinger, S. P., *J. Appl. Phys.* 46, 3800-3805 (1975).
9. Lau, Y. Y., Chu, K. R., Barnett, L. R., and Granatstein, V. L., *Int. J. Infrared and Millimeter Waves* 2, 373-393 (1981).
10. Chu, K. R., and Lin, A.T., *IEEE Trans. Plasma Sci.* 16, 90-104 (1988).
11. Davis, J. A., *Phys. Fluids B* 1, 663-669 (1989).
12. Sprangle, P., and Manheimer, W. M., *Phys. Fluids* 18, 224-230 (1975).
13. Sprangle, P., and Drobot, A., *IEEE Trans. Microwave Theory and Techniques* 25, 528-544 (1977).
14. Chu, K. R., Drobot, A. T., Granatstein, V. L., and Seftor, J. L., *IEEE Trans. Microwave Theory and Techniques* 27, 178-187 (1979).
15. Chu, K. R., Drobot, A. T., Szu, H. H., and Sprangle, P., *IEEE Trans. Microwave Theory and Techniques* 28, 313-317 (1980).
16. Seftor, J. L., Granatstein, V. L., Chu, K. R., Sprangle, P., and Read, M. E., *IEEE J. Quantum Electron.* 15, 848-853 (1979).
17. Barnett, L. R., Chu, K. R., Baird, J. M., Granatstein, V. L., and Drobot, A. T. in *Technical Digest of the International Electron Devices Meeting*. New York: IEEE, 1979, pp.164-167.
18. Barnett, L. R., Baird, J. M., Lau, Y. Y., Chu, K. R., and Granatstein, V .L., in *Technical Digest of the International Electron Devices Meeting,* New York: IEEE, 1980, pp. 314 - 317.
19. Lau, Y. Y., and Chu, K. R., *Int. J. Infrared and Millimeter Waves* 2, 415-425 (1981).

20. Chu, K. R., Lau, Y. Y., Barnett, L. R., and Granatstein, V. L., *IEEE Tran. Electron Devices* **28**, 866-871 (1981).
21. Barnett, L. R., Lau, Y. Y., Chu, K. R., and Granatstein, V. L., *IEEE Trans. Electron Devices* **28**, 872-875 (1981).
22. Symons, R. S., Jory, H. R., Hegji, S. J., in *Technical Digest of the International Electron Devices Meeting.* New York: IEEE, 1979, pp. 676-679.
23. Ferguson, P. E., and Symons, R. S., in *Technical Digest of the International Electron Devices Meeting,* New York: IEEE, 1980, pp. 310-313.
24. Symons, R. S., Jory, H. R., Hegji, S. J., and Ferguson, P. E., *IEEE Trans. Microwave Theory Tech.* **29**, 181-184 (1981).
25. Ferguson, P. E., Valier, G. and Symons, R. S., *IEEE Trans. Microwave Theory Tech.* **29**, 794-799 (1981).
26. Jory, H. R., private communication.
27. Granatstein, V. L., Read, M., and Barnett, L. R., in *Infrared and Millimeter Waves* **5**, K. J. Button, Ed. New York: Academic, 1984, pp. 267-304.
28. Symons, R. S., and Jory, H. R., in *Advances in Electronics and Electron Physics* **55**, L. Marton and C. Marton, Eds. New York: Academic, 1981, pp. 1-75.
29. Barnett, L. R., Chang, L. H., Chen, H. Y., Chu, K. R., Lau, Y. K., and Tu, C. C., *Phys. Rev. Lett.* **63**, 1062-1065 (1989).
30. Chu, K. R., Barnett, L. R., Lau, W. K., Chang, L. H., and Chen, H. Y., *IEEE Trans. Electron Devices* **37**, 1557-1560 (1990).
31. Chu, K. R., Barnett, L. R., Lau, W. K., Chang, L. H., Lin, A. T., and Lin, C. C., *Phys. Fluids B* **3**, 2403-2408 (1991).
32. Chu, K. R., Barnett, L. R., Lau, W. K., Chang, L. H., and Kou, C. S., in *Technical Digest of International Electron Devices Meeting.* New York: IEEE, 1990, pp.699-702.
33. Chu, K. R., Barnett, L. R., Chen, H. Y., Chen, S. H., Wang, Ch., Yeh, Y. S., Tsai, Y. C., Yang, T. T., and Dawn, T. Y., *Phys. Rev., Lett.* **74**, 1103-1106 (1995).
34. Chu, K. R., Chen, H. Y., Hung, C. L., Chang, T. H., Barnett, L. R., Chen, S. H., and Yang, T. T., *Phys. Rev. Lett.* **81**, 4760-4763 (1998).
35. Chu, K. R., Chen, H. Y., Hung, C. L., Chang, T. H., Barnett, L. R., Chen, S. H., Yang, T. T., and Dialetis, D., *IEEE Trans. Plasma. Sci.* **27**, 391-404 (1999).
36. Furuno, D. S., McDermott, D. B., Kou, C. S., Luhmann, Jr., N. C., and Vitello, P., *Phys. Rev. Lett.* **62**, 1314-1317 (1989).
37. Chong, C. K., McDermott, D. B., and Luhmann, Jr., N. C., *IEEE Trans. Plasma Sci.,* **26**, 500-507, (1998).
38. Wang, Q. S., McDermott, D. B., and Luhmann, Jr., N. C., *Phys. Rev. Lett.* **75**, 4322-4325 (1995).
39. Wang, Q. S., McDermott, D. B., and Luhmann, Jr., N. C., *IEEE Trans. Plasma Sci.* **24**, 700-706 (1996).
40. McDermott, D. B., Furuno, D. S., and Luhmann, Jr., N. C., *J. Appl. Phys.* **58**, 4501-4508 (1985).
41. Lin, A. T., Chu, K. R., Lin, C. C., Kou, C. S., McDermott, D. B., and Luhmann. Jr., N. C., *Int. J. Electron.* **72**, 873-885 (1992).
42. Kou, C. S., Wang, Q. S., McDermott, D. B., Lin, A. T., Chu, K. R., and Luhmann, Jr., N. C., *IEEE Trans. Plasma Sci.* **20**, 155-162 (1992).
43. Wang, Q. S., Kou, C. S., McDermott, D. B., Lin, A. T., Chu, K. R., and Luhmann, Jr., N. C., *IEEE Trans. Plasma Sci.* **20**, 163-169 (1992).
44. Park, G. S., Park, S. Y., Kyser, R. H., Ganguly, A. K., and Armstrong, C. M., in *Technical Digest of International Electron Device Meeting.* New York: IEEE, pp.779-781, 1991.
45. Park, G. S., Park, S. Y., Kyser, R. H., Armstrong, C. M., Ganguly, A. K., and Parker, R. K., *IEEE Trans. Plasma Sci.* **22**, 536-543 (1994).
46. Ganguly A. K., and Ahn, S., *Int. J. Electron.* **53**, 641-658 (1982).
47. Ganguly, A. K., and Ahn, S., *IEEE Trans. Electron Devices* **31**, 474-480 (1984).
48. Park, G. S., Choi, J. J., Park, S. Y., Armstrong, C. M., Ganguly, A. K., Kyser, R. H., and Parker, R. K., *Phys. Rev. Lett.* **74**, 2399-2402 (1995).
49. Gold, S. H., Kirkpatrick, D. A., Fliflet, A. W., McCowan, R. B., Kinkead, A. K., Hardesty, D. L., and Sucy, M., *J. Appl. Phys.* **69**, 6696-6698 (1991).

50. Fliflet, A. W., *Int. J. Electron.* **61**, 1049-1080 (1986).
51. Denisov, G. G., Bratman, V. L., Phelps, A. D. R., and Samsonov, S. V., *IEEE Trans. Plasma Sci.* **26**, 508-518 (1998).
52. Denisov, G. G., Bratman, V. L., Gross, A. W., He, W., Phelps, A. D. R., Ronald, K., Samsonov, S. V., and Whyte, C. G., *Phys. Rev. Lett.* **81**, 5680-5683 (1998).
53. Bratman, V. L., Gross, A. W., Denisov, G. G., He, W., Phelps, A. D. R., Ronald, K., Samsonov, S. V., Whyte, C. G., and Young, A. R., *Phys. Rev. Lett.* **84**, 2746-2749 (2000).
54. Lau, Y. Y., Chu, K. R., Barnett, L. R., and Granatstein, V. L., *Int. J. Infrared and Millimeter Waves* **2**, 395-413 (1981).

Interaction Circuits for High Average Power Gyro-TWTs Based on Monolithic Lossy Ceramics

Jeffrey P. Calame[*], Morag Garven[†], Bruce G. Danly[*], Baruch Levush[*], and Khanh T. Nguyen[§]

[*]Naval Research Laboratory, Washington, DC 20375, USA
[†]Omega-P, Inc., New Haven, CT 06520, USA
[§]KN Research, Inc., Silver Spring, MD 20905, USA

Abstract. Techniques for providing controlled loading of the TE_{01} operating mode of a 35 GHz, gyro-traveling wave tube (gyro-TWT) using monolithic, lossy ceramic structures are presented. The loading scheme, which also suppresses spurious backward-wave oscillations in the TE_{11}, TE_{21}, and TE_{02} modes, is based on a sequence of alternating ceramic cylindrical shells and metal rings that surround the beam tunnel. In this paper, we present design techniques for achieving optimal performance in this type of structure.

INTRODUCTION

The gyro-traveling wave tube (gyro-TWT) is an attractive candidate for use as the transmitter power amplifier in millimeter-wave radars. Gyro-TWTs are capable of much broader bandwidth than gyro-klystrons, while retaining the high power at high-frequency advantages inherent in the cyclotron maser mechanism. However, an important issue in gyro-TWT design is maintaining stability to backward-wave oscillation in the interaction space, while at the same time allowing for high average power operation in the desired mode.

A diagram of a typical gyro-TWT is shown in Fig. 1. The active medium is a spiraling electron beam immersed in an axial magnetic field. Traveling waves are launched into the cylindrical-cross section interaction space by an upstream-located input coupler. As the electromagnetic wave and the spiraling electron beam move down through the interaction space, the kinetic energy of the transverse (orbital) motion of the electron beam is transferred into the electromagnetic fields, creating rf amplification. The majority of the interaction space should exhibit a moderate amount of electromagnetic loss per unit length (measured without an electron beam), which helps stabilize the structure against oscillations [1]. Finally, the downstream side of the interaction space is completed with a short, unloaded cylindrical tunnel in which the final, highest powered portion of the amplification takes place.

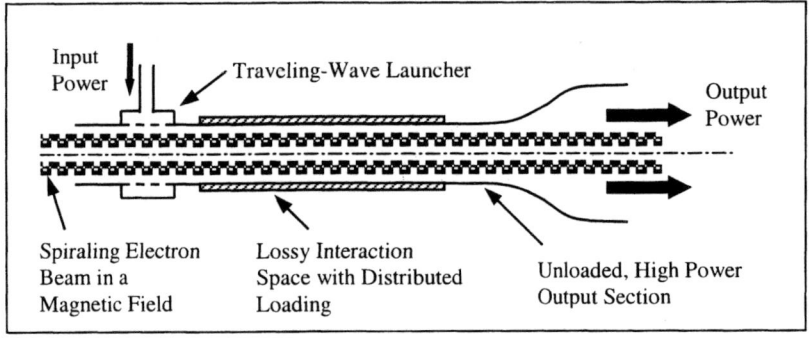

FIGURE 1. Schematic of a gyro-TWT.

We have been investigating techniques for loading the majority of the interaction space of a 35 GHz, TE_{01}-mode gyro-TWT using a lining of lossy ceramic cylindrical shells periodically interspaced with metal rings. In addition to allowing the adjustable suppression of the fundamental TE_{11} and TE_{21} and 2nd harmonic TE_{02} backward waves, the ceramic technique allows the desirable, controlled loading of the TE_{01} operating mode within the interaction space. High average power operation is possible by using high thermal conductivity lossy ceramics (AlN-SiC or BeO-SiC) that are in intimate thermal contact with the surrounding metal structure. This paper will describe the design techniques associated with the ceramic loading scheme, using the 35 GHz gyro-TWT as an example.

ILLUSTRATION OF LOADING REQUIREMENTS

Consider an example gyro-TWT designed for operation at 35 GHz in the TE_{01} mode, using a small-orbit spiraling beam produced by a magnetron injection gun. Furthermore, consider a nominal operating point of 65 kV beam voltage, 10 A beam current, a perpendicular-to-parallel velocity ratio α of 1.1, a magnetic field of 1.248 T, and an interaction space radius (wall radius) of 5.49 mm. In the absence of losses, and temporallily ignoring stability issues, linear gyro-TWT theory [2] predicts that the spatial growth rate of the operating TE_{01} mode to be about 3.0 dB/cm. Furthermore, nonlinear simulations with the MAGY code [3], in the absence of spurious reflections, indicate that peak powers of 150-200 kW and 3 dB bandwidths of 2-3 GHz are attainable with this simple design. Thus, the TE_{01} mode offers the potential for large output power at high frequency.

However, TE_{01}-mode devices are susceptable to backward wave oscillation in a number of other modes. Stability analysis with linear theory indicates very short critical oscillation lengths L_{crit} for the TE_{11} mode (L_{crit} = 7.46 cm, f_{osc} = 25.78 GHz), the TE_{21} mode (L_{crit} = 4.69 cm, f_{osc} = 28.87 GHz), and the TE_{02} mode (L_{crit} = 4.55 cm, f_{osc} = 61.30 GHz). These lengths are much smaller that the total circuit length required to create the 100-200 kW of saturated output power; hence, the example design will not work without a vigorous stabilization scheme. Furthermore, even though the TE_{01} operating mode is predicted to be stable at up to 14 A beam current, once realistic values of window reflections and reflections from the transition between the interaction space and the output waveguide are included, the start oscillation current drops to only 3 A, well below the operating current.

The addition of distributed losses into the interaction space can greatly mitigate these problems, as has been demonstrated in a TE_{11}-mode gyro-TWT by K.R. Chu and coworkers [1]. However, the TE_{01} device under consideration in the present paper has many more potentially unstable modes, so a careful analysis is required to identify the necessary degree of loading. A good starting point is an analysis using linear gyro-TWT theory that includes resistive wall losses. An illustration of the growth vs. frequency curves for the spurious TE_{11} mode, at a number of wall loadings, is shown in Fig. 2. As the wall losses increase, the growth curves evolve from a single solution in the zero loss case, shown in Fig 2(a), to a pair of curves that do not cross at 1.5 dB/cm loss, shown in Fig. 2(b), to a pair of curves that cross at 2.0 dB/cm, in Fig. 2(c), and finally to a pair of curves that form a cusp-like intersection, as in Fig 2(d) at 2.5 dB/cm. Any additional loss beyond that of Fig. 2(d) will cause the curves to separate at this location, at which time the system becomes stable to the backward wave mode. Thus, a set of calculations similar to those of Fig. 2, and a realization of the graphical significance of the breakpoint between solutions, can be used to determine the cold attenuation per unit length required to suppress a BWO. Roughly speaking, the cold spatial loss rate must exceed the BWO spatial growth rate that would have occurred in the absence of losses, at the oscillation frequency of the BWO. Thus, the system requires 2.5 dB/cm of loss at 25.2 GHz to suppress the TE_{11} mode. Similar

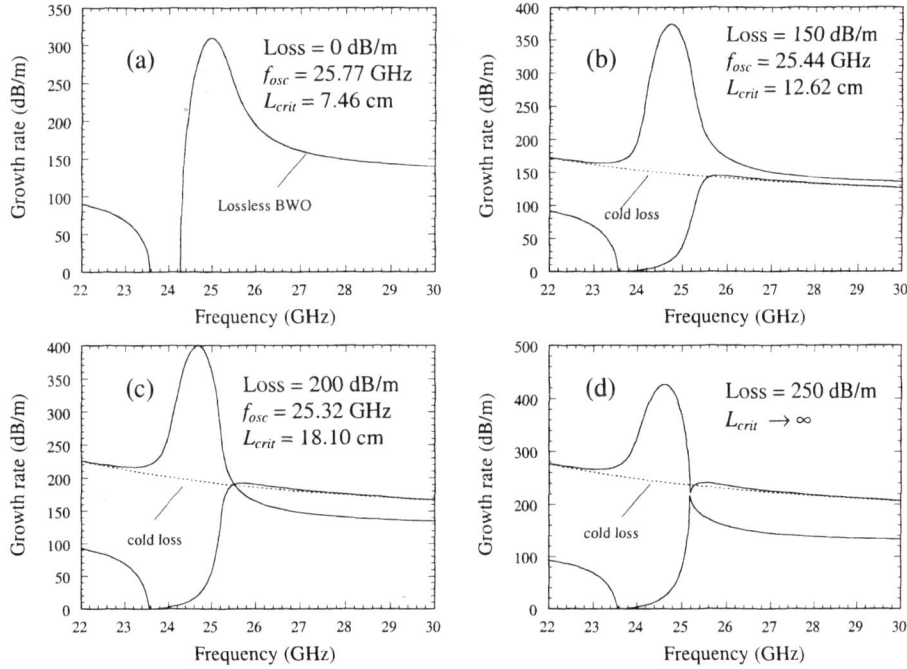

FIGURE 2. Spatial growth rate for the TE_{11} mode with various losses

calculations indicate that 4.5 dB/cm at 28.1 GHz is needed to stop the TE_{21} mode, and 4.2 dB/cm at 61.0 GHz is required to arrest the TE_{02} mode. To explore the distributed losses needed to ensure stability of the operating TE_{01} mode at 10 A, in a realistic situation including window and nonlinear uptaper reflections, a linear stability study was completed using the MAGY code. An attenuation of about 3 dB/cm at 35 GHz was found to be suitable.

DESIGN OF THE CERAMIC-LOADED INTERACTION SPACE

There are many ways to create the required levels of distributed loading in a gyro-TWT. Thin lossy coatings [1] and slotted waveguides backed up by lossy material [4] are two proven methods. However, a third option is to use linings of lossy ceramic materials interspaced with metal rings, as shown in Fig. 3. In this method the ceramics have a radial thickness Δr that is a significant fraction of a wavelength (of order $0.5\lambda/\sqrt{\varepsilon'}$, where λ is the free space operating wavelength and ε' is the real part of the material dielectric constant). There are a number of advantages to this monolithic ceramic approach. First, it builds strongly on existing, highly successful techniques for loading gyroklystron drift spaces and gyrotron beam tunnels. Second, considerable flexibility in independently controlling the loading of the operating mode versus the parasitic modes can be achieved by proper choice of the ceramic (*i.e.* a choice of dielectric properties) and by adjustment of the dimensions of the ceramic and metal. Dielectric loading can also lead to higher bandwidth, by virtue of a better match between the dispersion curves of the cyclotron waves and the electromagnetic waves supported by the dielectric-lined structure [5]. Finally, the use of high thermal conductivity ceramics leads to high average power capability, provided that appropriate fabrication techniques, such as ceramic-to-metal brazing or differential expansion shrink fitting, are employed.

The design procedure for the interaction structure is similar to the techniques used to stabilize gyroklystron drift tubes. The big difference in gyro-TWT design is the need to create a controlled, moderate loading of a normally propagating operating mode, as opposed to simply loading all modes as much as possible. The key step in achieving controlled loading is a careful selection of the radial thickness of the dielectric, which is illustrated in Fig. 4(a). In this plot the attenuation and propagation factor for the two lowest order TE_{0m} modes vs. dielectric thickness is shown, while keeping the inner radius of the interaction space a fixed at 5.49 mm. The frequency is 35 GHz and

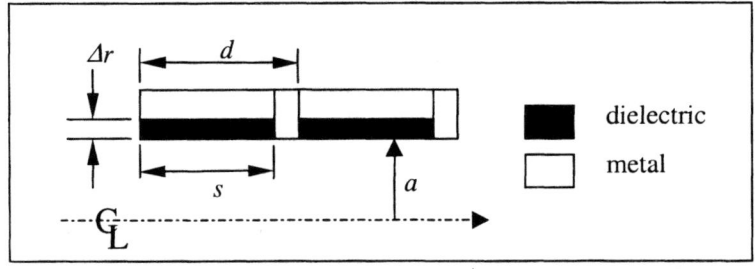

FIGURE 3. Diagram of the interaction space loading structure

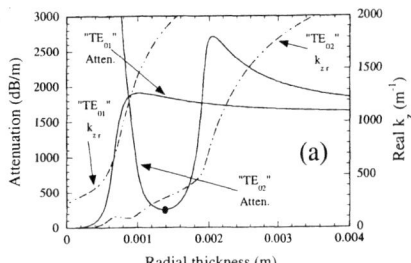 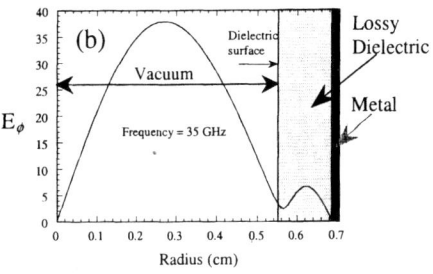

FIGURE 4. (a) Plot of TE0m mode behavior vs. dielectric thickness, (b) radial field profile of the second mode at a dielectric thickness of 1.4 mm

the complex dielectric permittivity is $\varepsilon^* = \varepsilon' - j\varepsilon'' = 11-2.2j$, which is an experimentally measured value for Ceradyne 80%AlN-20%SiC at room temperature.

The normally propagating mode in the absence of dielectric, $TE_{01(\Delta r=0)}$, has its electromagnetic fields rapidly pulled into the dielectric as the thickness Δr is increased towards 1 mm, leading to a peak in attenuation seen in Fig. 4(a). The level stays high for larger values of Δr. On the other hand, the next highest order TE_{0m} mode in the absence of dielectric, $TE_{02(\Delta r=0)}$, which does not normally propagate, has its attenuation dramatically reduced as the dielectric is added, and it eventually takes on propagating characteristics. Most interestingly, at a dielectric thickness near 1.4 mm the attenuation reaches a minimum (indicated by the dot in the figure), and the real propagation factor k_z is very similar to the propagation factor of the TE_{01} mode in the absence of dielectric. A radial field profile of the electric field for the 1.4 mm dielectric thickness case is shown in Fig. 4(b). Inside the vacuum ($0 \leq r \leq a$), the field profile is nearly identical to the $J_1(x_{01}'r/a)$ variation of the $TE_{01(\Delta r=0)}$ mode, which explains the similarity of the k_z values. The higher order radial mode structure is contained within the dielectric, allowing moderate losses. Furthermore, the similarity of the vacuum field profile to an ordinary TE_{01} mode also guarantees that the gyro-TWT interaction mechanism (i.e. the beam-wave interaction) will be similar to that of a resistively loaded circular waveguide, which is well understood and easily modeled. Thus, the "TE_{02}" mode in Fig. 4(a) will be referred to as the "TE_{01}-like" mode, which is the operating mode for the example gyro-TWT.

Lower values of attenuation per unit length, if required, are achieved by varying the other parameters of the geometry, such as the relative values of s and d, while striving to adjust Δr as to keep the attenuation vs. frequency curve near a local minimum. Maintaining this proximity to the local minimum is very important, since it will minimize the sensitivity of the attenuation to errors in dielectric thickness, due to machining tolerances, as well as lot-to-lot and temperature variations in the dielectric properties of the ceramic. Conversely, if higher values of attenuation are required, then the design process must be re-started with a more lossy ceramic material (*i.e.* a composite with a higher volume fraction of SiC).

The behavior of the optimized interaction space as a function of frequency is shown in Fig. 5 for the "TE_{01}-like" operating mode. The attenuation at 35 GHz is 3.44 dB/cm, as seen in Fig. 5(a). The dispersion diagram in Fig. 5(b) indicates a nearly

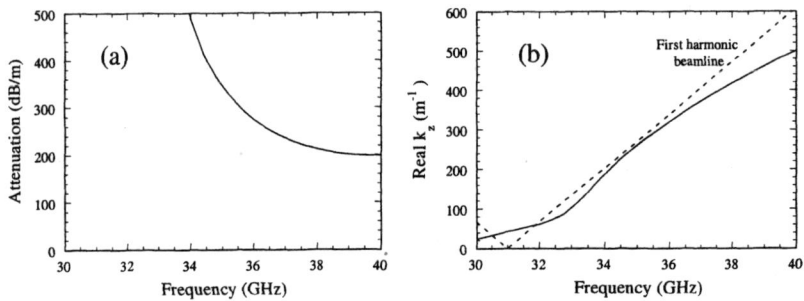

FIGURE 5. (a) Attenuation and (b) real k_z for the operating mode in the loaded interaction space.

grazing incidence with the first cyclotron harmonic beamline, as required for good gyro-TWT operation. Thus, the loading scheme is appropriate for the nominal TE_{01} operating mode. In addition, the structure exhibits attenuations well above the minimum values need to suppress the BWO modes. This is due to the fact that the interaction space was carefully designed to operate at a relative minimum in the attenuation vs. Δr curve for the operating mode, as in Fig. 4(a); such a condition is generally not met by the BWO modes, since their frequencies and field structures are very different. The attenuation presented to the TE_{11}-like, TE_{21}-like, and TE_{02}-like modes at the respective BWO frequencies are 3.04 dB/cm, 12.5 dB/cm, and 18.4 dB/cm, respectively. Thus, the interaction space will be stable.

SUMMARY OF ONGOING WORK

Experimental studies of a 35 GHz gyro-TWT using this interaction space are presently underway. Peak powers of 125-140 kW at about 49 dB gain have been achieved, and the stability is excellent. Since this is a proof-of-principle experiment, the research has been limited to a low duty cycle (below 0.5%) to eliminate the need for ceramic-to-metal brazing and water cooling, but a more robustly constructed version of the interaction space should be compatible with 10-20% duty cycle operation. At present the bandwidth is limited to about 1.1 GHz by the performance of the electron gun, which was made for a gyrotron and has an excessively high velocity spread $\Delta v_\perp / v_\perp$ of 7.5%. Simulations indicate that a gun with a spread of 3% will have 2.5-3 GHz of bandwidth.

ACKNOWLEDGEMENTS

This work was supported by the Office of Naval Research.

REFERENCES

1. Chu, K.R., Chen H.Y., Hung, C.L., *et. al.*, IEEE Trans. Plasma Sci. 27, 391-404 (1999).
2. Kou, C.S., Wang, Q.S., McDermott, D.B., *et. al.*, IEEE Trans. Plasma Sci. 20, 155-162 (1992).
3. Botton, M., Antonsen, T.M., Levush, B., et. al., IEEE Trans. Plasma Sci. 26, 882-892 (1998).
4. Wang, Q.S., McDermott, D.B., and Luhmann, N.C., IEEE Trans. Plasma Sci. 24, 700-706 (1996).
5. Leou, K.C., McDermott, D.B., and Luhmann, N.C., IEEE Trans. Plasma Sci. 20, 188-196 (1992).

140 kW W-Band Heavily Loaded TE01 Gyro-TWT Amplifier

D.B. McDermott, H.H. Song, Y. Hirata, A.T. Lin[1], T.H. Chang[2], H.L. Hsu[2], K.R. Chu[2], and N.C. Luhmann, Jr.

Department of Applied Science, University of California, Davis
[1] *Department of Physics, UCLA, Los Angeles, CA*
[2] *Department of Physics, NTHU, Hsinchu, Taiwan*

Abstract. A high power gyro-TWT operating in the low-loss TE_{01} mode has been constructed at UCD that is driven by a 100 kV, 5 A, $v_\perp/v_z=1.0$ MIG electron beam with $\Delta v_z/v_z=5\%$. The amplifier is predicted by a large-signal simulation code to generate 140 kW at 94 GHz with 28% efficiency, 50 dB saturated gain and 5% bandwidth.

GYRO-TWT DESIGN

A broadband 140 kW, 94 GHz amplifier has been built at UCD that operates in the low loss TE_{01} mode [1]. Because a large axial velocity is beneficial for stability and wide bandwidth, a 100 kV, 5 A beam with a velocity ratio of $v_\perp/v_z=1$ is chosen. To reduce the interception of electrons by the wall, the guiding center of the beam is $r_c/r_w=0.45$, which is slightly inside of the mode maximum.

A self-consistent code has been employed to evaluate the large-signal characteristics of the amplifier. The parameters are given in Table I. For an axial velocity spread of $\Delta v_z/v_z=5\%$, the predicted peak power is 140 kW with an efficiency of 28% and a saturated bandwidth of 5%, as shown in Fig. 1. The large signal gain is 50 dB, even though the circuit has been heavily loaded. The loss at the center frequency of 93 GHz is 90 dB.

To stabilize the amplifier, the walls are coated with Aquadag, a lossy carbon colloid, a technique recently employed to operate a high performance Ka-band gyro-TWT [2]. Figure 2 shows that an insertion loss of 90 dB has been measured at 93 GHz in the circuit with an unloaded cutoff of 91 GHz. So that the wave is not damped in the high power region, there is no loss in the final 2.5 cm of the circuit and the loss in the previous 1 cm is linearly tapered. This would allow cw operation at 140 kW with 50 W/cm^2 wall loading.

A coaxial-filter input coupler has been designed with HFSS with a 1 dB insertion loss over the 5% bandwidth. As shown in Fig. 3, the input signal is injected from conventional rectangular waveguide into the coaxial TE_{51} mode of a coaxial cavity and then into the desired TE_{01} mode of the cylindrical interaction waveguide within the inner coax through five slots in the wall. A photograph of the input coupler is shown in Fig. 4. As shown in Fig. 5, the measurements of the input coupler are in fair agreement with HFSS. The bandwidth is 4% and the coupling is 2 dB. An output coax coupler to monitor the peak power was measured with a coupling of 10 dB and 5% bandwidth.

The entire circuit has been fabricated. The MIG electron gun with a predicted axial velocity spread of 5% has been built. The edges of the emitting strip are coated with Molybdenum to suppress edge emission. A 100 W coupled-cavity TWT will drive the gyro-TWT into saturation and a 1 kW EIA is available as a back-up. The tests are being performed in a 50 kG superconducting magnet and have begun.

ACKNOWLEDGMENTS

This work has been supported by AFOSR under Grants F49620-99-1-0297 (MURI MVE) and F49620-00-1-0339.

REFERENCES

1. L.R. Barnett, et al., *IEDM*, 314 (1980).
2. K.R. Chu, et al., *Phys. Rev. Lett.* **81**, 4760 (1998).

TABLE I. Design parameters of the heavily loaded TE_{01} gyro-TWT amplifier.

Voltage	100 kV
Current	5 A
$\alpha = v_\perp/v_z$	1.0
$\Delta v_z/v_z$	5%
Magnetic Field, B_0	35.6 kG
B_0 / B_g	0.995
Cutoff Frequency	91.0 GHz
Wall Resistivity	70,000 ρ_{Cu}
Guiding Center Radius, r_c	0.45 r_w
Circuit Radius, r_w	0.201 cm
Copper Circuit Length	2.5 cm
Total Circuit Length	14.5 cm

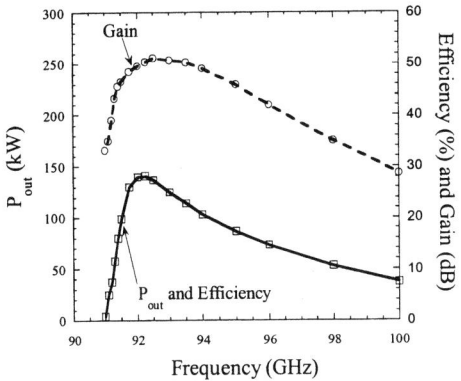

FIGURE 1. Predicted saturated bandwidth of the output power, efficiency and gain [Table I].

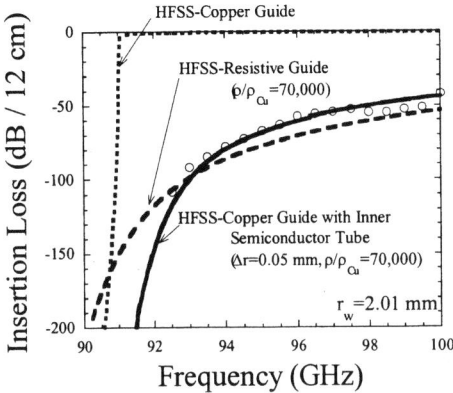

FIGURE 2. Dependence of insertion loss on frequency for TE_{01} mode through the loaded 12 cm circuit from measurement (circles) and HFSS (unbroken curve), where a semiconductor tube with thickness of 0.05 mm and resistivity 70,000 times copper is within the copper guide. Also shown is the loss from HFSS for metallic guide with a resistivity of copper (dotted curve) and 70,000 times copper (dashed curve).

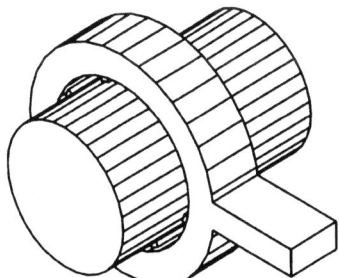

FIGURE 3. Schematic of the coax coupler showing the rectangular input guide splitting in a tee and then wrapping around the interaction circuit.

FIGURE 4. Photograph of the input coax coupler showing the rectangular input guide joining the outer coax in a tee junction, the inner interaction circuit, and five interconnecting slots.

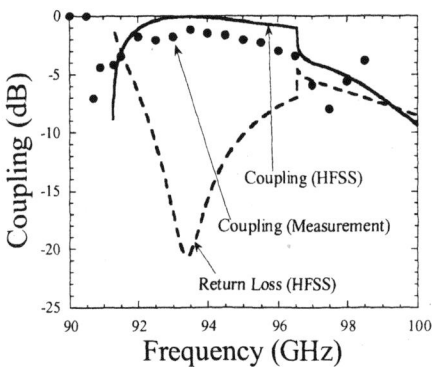

FIGURE 5. Bandwidth of the input coax coupler from HFSS and measurement.

New opportunities in vacuum electronics using photonic band gap structures

J. R. Sirigiri, C. Chen, M. A. Shapiro, E. I. Smirnova, and R. J. Temkin

*Plasma Science and Fusion Center, Massachusetts Institute of Technology
NW16, 167 Albany Street, Cambridge, MA 02139*

Abstract. The advantages of using photonic band gap structures in vacuum electron devices are discussed. Their excellent mode selective properties permit them to be used as overmoded interaction structures in microwave tubes for either high power (>10 MW) or extremely high frequency (>100 GHz) operation. Two proof-of-principle experiments recently conducted at MIT namely, a 17 GHz PBG resonator with a TM_{010}-like mode for potential applications in accelerators and conventional slow wave devices and a 140 GHz highly overmoded gyrotron with a TE_{041}-like mode PBG resonator are presented. A brief outline of the theory of PBG structures and some of their additional applications in microwave vacuum electronics are also discussed.

INTRODUCTION

The extension of power output of microwave tubes to levels greater than 100's of MW and the operating frequency into the millimeter and sub-millimeter wave regime is limited by the size of the interaction structure. Higher power operation needs a larger interaction structure for better thermal dissipation capability while higher frequency operation is restricted by the rapid miniaturization of the size of the interaction structure with increasing frequency, which drastically increases the complexity of fabrication. Both these problems can be mitigated by the use of overmoded interaction structures with transverse dimensions much larger than the operating wavelength but such structures are plagued by mode competition. Clearly, an overmoded interaction structure with mode-selective is desired.

Photonic band gap (PBG) structures are very promising for use as overmoded interaction structures in a variety of microwave tubes spanning both the conventional slow-wave devices and the fast-wave devices such as the gyrotron. PBG structures are 2D or 3D lattices of metal and/or dielectric elements. Their spatial periodicity renders them with frequency selective properties, which allow them to be used as mode-selective structures.

We present a brief theory of PBG structures to lay the foundation for designing mode-selective interaction structures. Two recent proof-of-principle experiments [1-2] at MIT using PBG structures for high power microwave sources and accelerators will be described followed by a discussion on the additional applications of PBG structures for other microwave tubes.

THEORY OF PBG STRUCTURES

PBG structures can be two- or three- dimensional periodic lattices of metal and/or dielectric elements [3-4]. Either square or triangular lattices are possible and their analysis is essentially similar. Triangular lattices find more favor while designing structures with cylindrical symmetry because removal of a few elements from a triangular lattice would leave a more cylindrical defect (hole) than in the case of a square lattice.

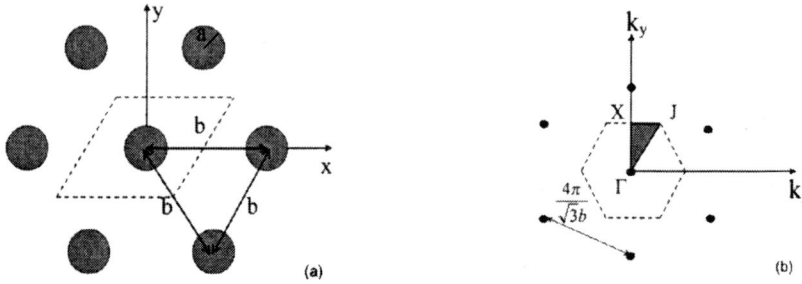

FIGURE 1. (a) Cross-section of a PBG structure made of a triangular lattice of perfectly conducting metal cylinders. The dotted parallelogram is the boundary of the unit cell used in the finite-difference analysis of the structure. (b) Reciprocal lattice and Brillouin zone for the lattice. The shaded region is the irreducible Brillouin zone.

We consider a triangular lattice of metal posts as shown in Fig. 1(a). The propagation characteristics of electromagnetic waves through this lattice can be solved using a finite-difference algorithm described in [5]. The system can be solved for two different kinds of modes namely, electric field transverse to the rods (TE) and electric field parallel to the rods (TM). The dispersion diagram for the TE and TM waves in such a structure is shown in Fig. 2.

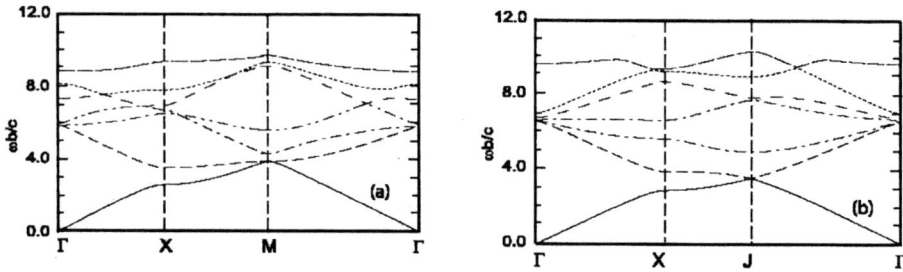

FIGURE 2. Dispersion diagram for a triangular lattice PBG structure for TM modes (a) and TE modes (b) for the filling fraction (a/b =0.2).

For the filling fraction of a/b = 0.2 there are no gaps in the dispersion diagram as shown in Fig. 2 however, by changing the filling fraction one can selectively open up a global band gap in the dispersion diagram. If for a range of frequencies there is no corresponding value of the wave vector k as one moves along the boundary of the irreducible Brillouin zone, then the structure is said to have a global band gap in that range of frequencies. For a frequency lying in the global band gap the structure behaves like an opaque wall while at other frequencies the structure is partially transmitting. The global band gaps for both the TM and TE modes in a triangular lattice PBG structure are shown in Fig. 3.

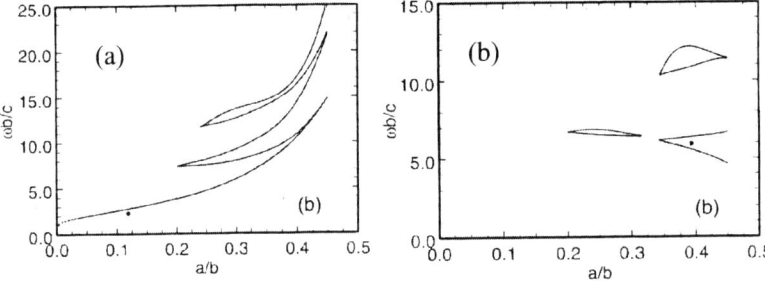

FIGURE 3. Global frequency band gaps for TM modes (a) and TE modes (b) in a triangular lattice PBG structure for various values of the filling fraction. The dot in (a) shows the operating point of the 17 GHz PBG resonator and the dot in (b) shows the operating point of the 140 GHz PBG resonator to be described in the next section.

EXPERIMENTS

Two recent experiments conducted at MIT to demonstrate the potential of PBG structures in microwave vacuum electronics are described in this section. These experiments were chosen to demonstrate the advantages of using PBG interaction structures for accelerators, slow-wave devices such as klystrons and fast wave devices such as gyrotrons.

17 GHz PBG Resonator

An oversized TM_{010}-like mode PBG resonator was designed, built and tested at 17 GHz for its mode-selective properties. Such oversized resonators will have better power handling capability and less stringent requirements on fabrication. They can be used in slow-wave microwave devices such as klystrons where their mode-selective properties can be used to suppress harmonic modes in the interaction structure. The suppression of non-operating modes in the PBG resonator will prove to be useful in reducing the wakefields in a charged particle accelerator.

The resonator was made of a triangular lattice of metal rods from which one central rod was omitted to form as defect as shown in Fig. 4 (a). The lattice parameters were chosen such that there is band gap at 17 GHz the operating frequency while at higher frequencies the lattice would act like a semi-transparent wall to suppress higher order modes. The operating point of the resonator is shown as a dot in Fig 3 (a). A TM_{010}-like mode can now be confined in the defect as shown in HFSS (Ansoft high frequency structure simulator) simulations in Fig. 4 (b). Some of the design and operating parameters of the resonator are listed in Table 1.

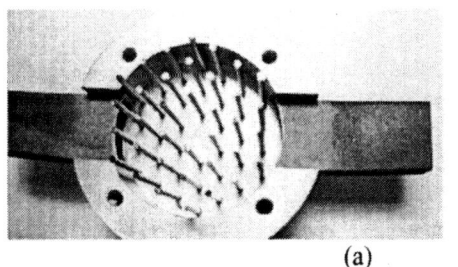
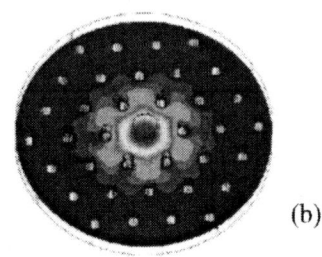

FIGURE 4. Picture of the 17 GHz PBG resonator formed by a defect (missing rod in the center) in a triangular lattice **(a)** and HFSS simulations showing a TM_{010}-like eigenmode confined in the defect **(b)**.

TABLE 1. Parameters of the 17 GHz PBG resonator

Lattice vector (b)	0.64 cm
Rod radius (a)	0.079 cm
Resonator radius	2.15 cm
Resonator length	0.787 cm
Eigenfrequency	17.32 GHz
Ohmic Q (theoretical)	5200
Shunt impedance	2.1 MΩ/cm

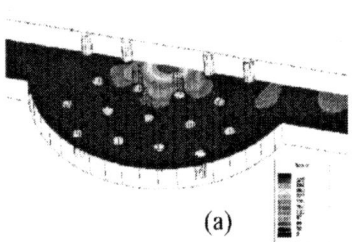
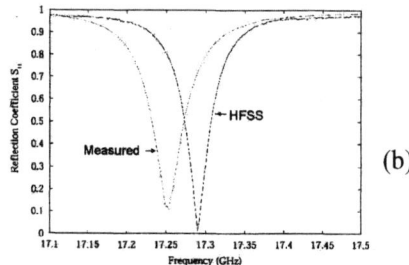

FIGURE 5. HFSS simulations showing the coupling of the coupling of power into the PBG resonator from an external waveguide excited in the TE_{10} mode **(a)** and experimental measurement of the coupling characteristics **(b)**.

The ease of coupling power into and out of the resonator is a very important for its suitability in microwave sources. Power can be coupled in and out of the PBG structure by using a very simple scheme shown in Fig. 5 (a). A few rods can be removed from the outer rows to allow the coupling of power from an external waveguide. To achieve critical coupling in the experiment 2 rods were removed from the fourth row, 4 rods were removed from the third row and two rods were partially withdrawn from the second row. The experimentally measured S_{11} of the resonator in the above configuration is shown in Fig. 5 (b). The experimentally measured Ohmic Q was 500. Another coupling scheme where the external waveguide is rotated by 90 degrees in the cross-sectional plane of the resonator was also tried with similar results

[2]. Since there is no iris through which power is coupled into the structure the problems associated with enhanced field intensity at the iris leading to breakdown in conventional pillbox cavities can be avoided.

140 GHz PBG Resonator Gyrotron

Gyrotrons are fast wave devices that use large overmoded resonators for interaction between a mildly relativistic electron beam and a higher order TE resonator mode. Naturally, the interaction in such overmoded resonators is beset with mode-competition, which prevents access to the high efficiency operating parameters of the design mode. An overmoded yet mode-selective resonator by suppressing the mode-competition would allow operation of the device in a very higher order mode which is essential for using large dimension resonators at small wavelengths (e.g. sub-millimeter waves).

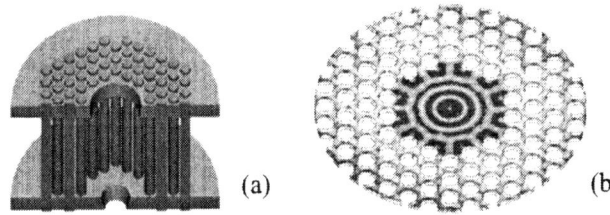

FIGURE 6. A section of the CAD drawing of the PBG resonator used in the gyrotron experiment (a) and HFSS simulations of the TE_{041} like eigenmode of the PBG resonator at 140 GHz (b).

A PBG resonator was designed for a 140 GHz gyrotron experiment with a triangular lattice of metal rods (Fig. 6). Nineteen inner rods were removed from the lattice so as to create a hole just big enough to localize the TE_{041}-like mode of a conventional cylindrical resonator. The actual resonator in the experiment was formed of 102 copper rods of 1.59 mm diameter with a spacing of 2.03 mm between the rods (center to center). The rods were held in position by two copper endplates. Radiation that passes through the array propagates out and is not reflected back into the lattice, permitting a strong single mode operation.

The nominal operating voltage of the gyrotron is 68 kV at a beam current of 5A with a beam velocity pitch factor (ratio of transverse to longitudinal velocity) of 1.2. In order to test the PBG gyrotron oscillator for mode selectivity, the device was operated over the magnetic field range of 4.1 o 5.8 T corresponding to an equal range in frequency tuning of about 30 %. The variation of output power with the magnetic field, the most vital indicator of the mode selectivity of the resonator is shown in Fig. 7. The mode with the operating frequency of 140.05 GHz is the only strong mode emanating from the resonator. This result is a direct confirmation of the mode selectivity of the PBG resonator. A conventional cylindrical resonator operating over the same range of magnetic tuning was in this experiment, would face competition from at least 7 other modes. One of those competing modes, the TE_{241} mode, which is nearly degenerate with the TE_{041} mode, severely reduces its operating range and efficiency is absent in the PBG resonator. The maximum power recorded in the TE_{041} mode for the operating voltage and current used for the magnetic fields scan is about 16 kW. Operation at a different voltage and current produced up to 25 kW at an efficiency of 7 %.

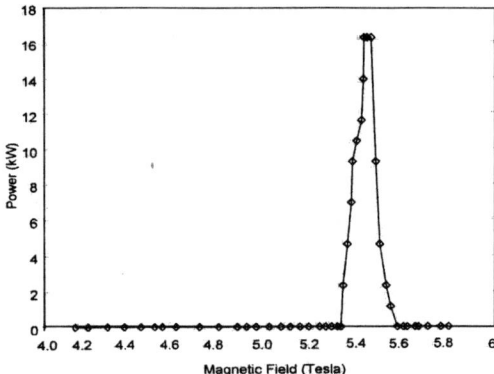

FIGURE 7. Variation of the output power of the PBG gyrotron with the main magnetic field.

At magnetic field values away from the operating mode, the gyrotron oscillator has weak emission in other modes, which can be excited in the output waveguide structure. Measurements based on a calibrated WR-8 video detector diode confirmed that the power in the points shown as 0 kW was everywhere less than 100W, which is at least 22 dB below the main mode.

For convenience, in these first experiments, we have used a flat plate with a hole for output coupling. HFSS simulations predict an ohmic Q factor of about 13500 (comparable to that of a equivalent cylindrical cavity) and a diffractive Q factor of about 16000 for this PBG cavity. This implies that more than half of the generated power is trapped inside the cavity leading to a low efficiency of the device. In the future, we plan to test PBG cavities with optimized output coupling including transverse coupling to reduce the diffraction Q factor. The transverse coupling in to the 17 GHz resonator has yielded excellent results.

DISCUSSION

Interaction structures based on PBG structures appear to be very promising for building overmoded microwave sources and amplifiers either at high power or at very high frequencies. The absence of mode competition in a PBG interaction structure allows operation at higher order modes thus reducing the average power density in the interaction structure for high power operation. Overmoded interaction structures at sub-millimeter wavelengths can be easily fabricated using conventional techniques.

At high average power, however, the rods of the PBG structure may not be able to dissipate ohmic losses as effectively as the smooth walls of conventional cylindrical cavities. This can be mitigated by using thicker rods and by cooling the rods with water flowing through channels in the center of each rod. The PBG structures would be able to handle high peak power levels, and would be particularly well suited to high peak power, moderate average power level amplifiers. They are also very attractive for use as the buncher cavities in amplifiers over a wide range of frequencies and any power level. A PBG interaction structure for a traveling wave amplifier can be designed such that the backward wave frequency is forced to lie in the pass band of the PBG structure thus preventing the confinement of the backward wave mode. This will reduce the vulnerability of the amplifier to backward wave oscillations at high operating currents. This technique may prove to be particularly useful in the current efforts to build a high average power W-band gyro-TWT. PBG structures hold a lot of promise for building overmoded conventional microwave tubes like TWTs, klystrons, etc. at frequencies well above 100 GHz.

ACKNOWLEDGMENTS

The authors wish to thank Ivan Mastovsky and William Mulligan for their help in running the experiments. This work was supported by the U.S. Dept. of Defense under the MURI (MVE) 1999; the U.S. Dept. of Energy under the Fusion Energy Sciences Program and the High Energy Physics Contract No. DE-FG02-91ER40648 and by AFOSR grant no. F49620-1-000007.

REFERENCES

1. Sirigiri J. R., Kreischer K. E., Machuzak J., Mastovsky I., Shapiro M. A., and Temkin R. J., *Phys. Rev. Lett.*, **86**, 5628-5631 (2001).
2. Shapiro M. A., Brown W. J., Mastovsky I., Sirigiri J. R., and Temkin R. J., *Phys. Rev. Special Topics- Accelerators and Beams*, 4, 042001 (2001).
3. Yablonovitch E., Gmitter T. J., and Leung K. M., *Phys. Rev. Lett.*, **67**, 2295-2298 (1997).
4. J. D. Joannopoulus, R. D. Meade and J. N. Winn, in *Photonic Crystals: Molding the Flow of Light*, Princeton: Princeton University Press, 1995.
5. Smirnova E. I, Chen C., Shapiro M. A., Sirigiri J. R., Temkin R. J., Submitted to *J. of Appl. Phys.* (2001).

Progress Toward a Gigawatt-Class Annular Beam Klystron with a Thermionic Electron Gun

M. Fazio[*], B. Carlsten[*], J. Farnham[*], K. Habiger[*], W. Haynes[*], J. Myers[*],
E. Nelson[*], J. Smith[*], B. Arfin[†], A. Haase[†], G. Scheitrum[†]

[*]*Los Alamos National Laboratory,*
MS-H851, Los Alamos, NM 87545, USA
[†]*Stanford Linear Accelerator Center*
2575 Sand Hill Rd. MS-33 Menlo Park, CA 94025, USA

Abstract. In an effort to reach the gigawatt power level in the microsecond pulse length regime Los Alamos, in collaboration with SLAC, is developing an annular beam klystron (ABK) with a thermionic electron gun. We hope to address the causes of pulse shortening in very high peak power tubes by building a "hard-vacuum" tube in the 10^{-10} Torr range with a thermionic electron gun producing a constant impedance electron beam. The ABK has been designed to operate at 5 Hz pulse repetition frequency to allow for RF conditioning. The electron gun has a magnetron injection gun configuration and uses a dispenser cathode running at 1100 °C to produce a 4 kA electron beam at 800 kV. The cathode is designed to run in the temperature-limited mode to help maintain beam stability in the gun. The beam-stick consisting of the electron gun, an input cavity, an idler cavity, and drift tube, and the collector has been designed collaboratively, fabricated at SLAC, then shipped to Los Alamos for testing. On the test stand at Los Alamos a low voltage emission test was performed, but unfortunately as we prepared for high voltage testing a problem with the cathode heater was encountered that prevented the cathode from reaching a high enough temperature for electron emission. A post-mortem examination will be done shortly to determine the exact cause of the heater failure. The RF design has been proceeding and is almost complete. The output cavity presents a challenging design problem in trying to efficiently extract energy from the low impedance beam while maintaining a gap voltage low enough to avoid breakdown and a Q high enough to maintain mode purity. In the next iteration, the ABK will have a new cathode assembly installed along with the remainder of the RF circuit. This paper will discuss the electron gun and the design of the RF circuit along with a report on the status of the work.

INTRODUCTION

High power RF sources simultaneously capable of gigawatt power levels and pulse lengths of a microsecond or more have so far eluded researchers in this area. The effort described in this paper attempts to reach this level of performance by taking as a starting point the enormous body of knowledge and experience on conventional high power klystron design, engineering, and fabrication. To this knowledge base we apply several new ideas including the use of an annular electron beam, advanced thermionic electron gun geometry, and new techniques for spatially modulating the electron beam so as to achieve a reasonably high efficiency for the conversion of modulated beam power to microwave power. Unlike most HPM sources, this annular beam klystron is designed for repetitively pulsed operation. Repetitive pulse capability allows for RF conditioning the tube up to high power in the same manner as commercial high power klystrons are conditioned in order to reach their design power level. We expect that the combination of extremely good vacuum, thermionic gun, and RF conditioning will allow microsecond performance at the gigawatt power level without incurring the familiar pulse shortening phenomenon.

ELECTRON GUN

The target conversion efficiency for the ABK is 35%. Therefore, to produce 1 GW of output power requires a beam power of about 3 GW. An annular beam is required to keep the beam potential depression as low as possible; otherwise the efficiency suffers because the kinetic energy is reduced by the amount converted to potential energy due to the high space charge. An operating voltage of 800 kV was chosen because of its practicality with respect to the modulator design. The current was set to be 4 kA resulting in an electron beam power of 3.2 GW. The first challenge is to produce a stable, minimal-halo, 4 kA electron beam. Because we wanted to build a "hard vacuum", sealed tube, we chose to use a thermionic gun based on standard dispenser cathode technology. A Pierce-type gun configuration was examined and produced an acceptable beam but the required diameter for the annular cathode and the surrounding ceramic insulator was considerably larger than the bore size for the focusing magnet. Consequently, we adopted a magnetron injection gun geometry, which allowed the emitting surface to be a tapered cylinder. The emission area was determined by the need to run at 20 A/cm^2 emission current density so as to have a cathode lifetime long enough to RF condition and thoroughly test the tube.

For beam stability in the MIG configuration, we needed to run the cathode in the temperature-limited mode. The steady state, finite-element gun code DEMEOS enabled rapid design iterations to ascertain the optimal cathode and focusing electrode shapes. The time-dependent behavior of the intense electron beam was modeled using the particle-in-cell code ISIS to look at the issue of beam stability, a major concern with an annular beam. ISIS is a generalized particle-in-cell code that pushes particles in the electromagnetic field in 2-1/2 dimensions. ISIS uses a large number of particles and a body-fitted coordinate system with a robust charge-conserving current algorithm and is therefore quite accurate. ISIS handles both temperature limited and space charge limited emission. ISIS computed a static solution for the space charge limited case, which was very challenging for DEMEOS. The DEMEOS final design is shown in Fig. 1. The emission surface is slightly angled to the horizontal and the beam undergoes a slight compression. The beam parameters are 800 kV and 4 kA giving a perveance of 5.6 μP. The drift tube radius is 4.25 cm and the beam filling-factor is 0.75. The axial magnetic field varies from about 1.2 kG at the cathode to a maximum value of about 5 kG.

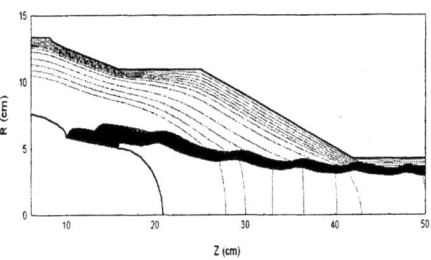

FIGURE 1. DEMEOS output showing electron gun cathode and anode configuration and beam.

The cathode assembly of the MIG gun is shown in Fig. 2. Measurements showed that temperature uniformity of the cathode is within 5°C at the 1100°C operating temperature. Since the cathode is being operated in the temperature-limited mode, excellent temperature uniformity across the surface is important.

BEAMSTICK

Because achieving a stable, high quality, electron beam is so important to the operation of the ABK, we wanted to demonstrate that an acceptable electron beam could be produced before building the complete tube. A beamstick was designed and built by SLAC consisting of the electron gun, two RF cavities, and the beam collector.

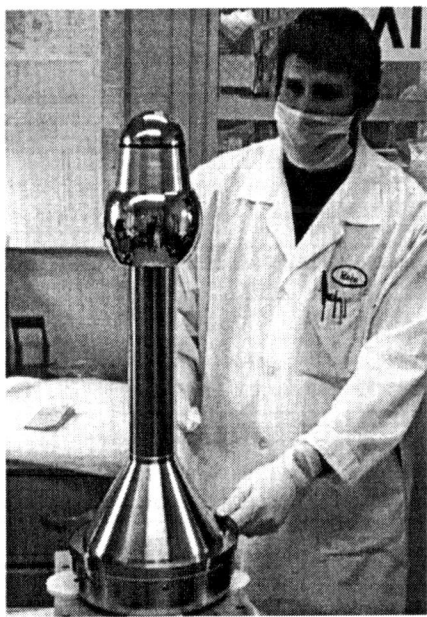

FIGURE 2. Magnetron injection gun cathode assembly.

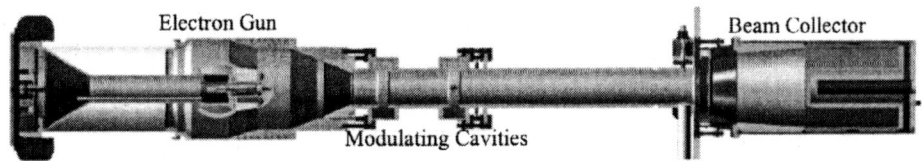

FIGURE 3. Annular beam klystron beam stick assembly consisting of magnetron injection gun, two RF modulating cavities, drift tube, and beam collector.

A cross-section of the beamstick is shown in Fig. 3. Both RF cavities have coupling loops. The loops will allow us to excite the beam in one cavity and examine the degree of modulation induced on the beam at the position of the second cavity. This data would provide useful information on the gain per cavity we can expect. The cavities will also be used as a beam diagnostic to detect beam halo. By varying the magnetic field profile we can tune the beam diameter. At the same time we can examine the field in the cavities and observe where the field becomes unstable due to beam interception and RF breakdown. This diagnostic should give us a feel for the importance of the beam halo problem, and for how close we can transport the beam to the drift tube wall.

The beamstick assembly is a furnace-brazed unit that undergoes a high temperature vacuum bake-out. The beamstick was baked out for several weeks and the cathode heater filament was run continuously at full power (2.5 kW) for about 3 weeks as the vacuum improved. After the vacuum reached an acceptable level the beamstick was shipped to Los Alamos and installed on the Banshee modulator for high power testing.

A low-voltage (100 V) cathode emission test was conducted. Without a magnetic field applied, the perveance was calculated to be 27 µP, which means that at 100 V the cathode emission current should be 27 mA. The experimental measurement was 27 mA indicating that the cathode geometry was performing as the codes predicted.

In preparing the beamstick for high power testing, a check of the cathode heater filament discovered a problem that had developed with the heater resistance. It had become a factor of four too low, probably due to a mechanical failure within the heater winding assembly during shipping from SLAC to Los Alamos. The heater windings were not potted but instead were supported by mechanical spacers. The stringent requirement on temperature uniformity imposed by temperature-limited operation coupled with uncertainty about the cathode assembly thermal design led us to use a heater design that we could modify. This desire for flexibility eliminated the possibility of potting the heater, which would have resulted in a more mechanically robust design. As it turned out, the thermal design was very good so the next heater will be potted. The low resistance heater problem was temporarily solved by thermally shocking the heater by applying twice the rated heater current (100 A) to it. The heater/cathode assembly was run for several hours at full power to make sure the tube was thoroughly outgassed. Later, as the cathode heater was subsequently run up to its design operating value of 55 A for high power testing, erratic behavior of the heater resistance was again encountered and the heater reverted to its original pre-thermal-shock resistance. Thermal shocking as before, high voltage pulsing, or mechanical shocking could not correct this low resistance condition. At this point, the testing was aborted because the cathode could not reach a high enough temperature to achieve electron emission. The tube was shipped back to SLAC for analysis and rebuild.

RF CIRCUIT DESIGN

The design of the RF circuit is driven primarily by the space charge dominated beam physics. The combination of the unusually high potential depression of the beam and the short overall tube length makes the RF circuit design complex. For conventional klystrons, the gain cavities are typically spaced about one-quarter reduced plasma wavelength apart. For the ABK tube, one-quarter of the reduced plasma wavelength is about 1 m, almost the entire length of the tube. Since we must place our cavities much closer together, we can recover the gain per cavity by driving each cavity harder (to increase the amount of harmonic current at the location of the next cavity), but that will increase the energy spread on the beam and lead to virtual cathode formation producing reflected electrons, low efficiency, instabilities, and RF breakdown. The solution is to adiabatically compress the beam with extra cavities to help minimize the induced energy spread.

Using a total of seven cavities, we can increase the harmonic current to about 75% of the average beam current at the position of the output cavity, while keeping cavity gap voltages to 300 kV or less. The penultimate cavity is capacitively detuned in order to compress the energy spread, with only a small penalty in induced current at the location of the output cavity. In a conventional klystron the penultimate cavity is inductively tuned to further increase the beam bunching. For our parameters with high space charge, the energy spread induced before the penultimate cavity is larger than the voltage of the penultimate cavity, and the beam is already well bunched by the time it arrives at the location of the penultimate cavity. While intense beams are relatively easy to bunch, the challenge lies in keeping the energy spread of the beam to a minimum.

A new variant of the ISIS particle-in-cell code was used for simulating the ABK circuit design. This new code, TUBE, written for modeling traveling-wave interactions, was modified to include standing-wave cavities. TUBE was used rather than ISIS because less dense particle injection is needed for modeling RF tubes than for general plasma problems. The key element of ISIS that needed to be preserved for this application is the unique charge conserving current algorithm, that eliminates the extra Poisson solve typically needed for charge conservation.

In Figures 4 and 5 we can compare the effects of inductive and capacitive tuning of the penultimate cavity. The induced current at the location of the output cavity at $z = 1.2$ m is 65% with the inductive tuning and 56% with the capacitive tuning. This decrease in the induced current is more than compensated for by the drastic decrease

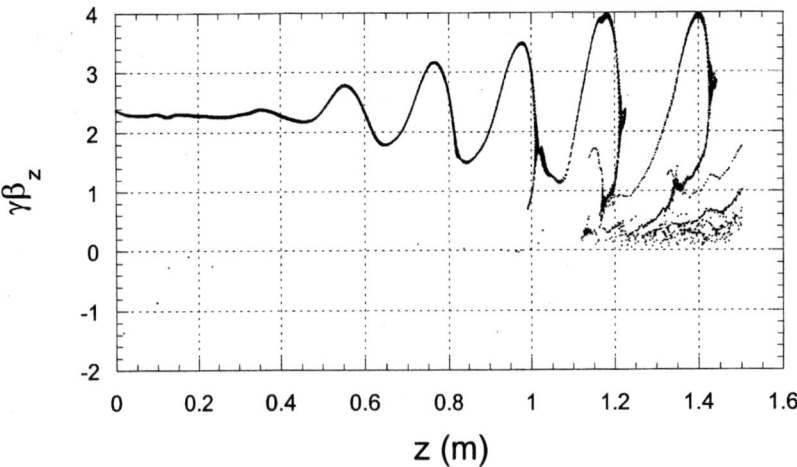

FIGURE 4. Beam energy versus axial position z along the ABK with inductively tuned penultimate cavity. Output cavity location is at z=1.2 m. Note large energy spread at z = 1.2 m and the formation of a virtual cathode.

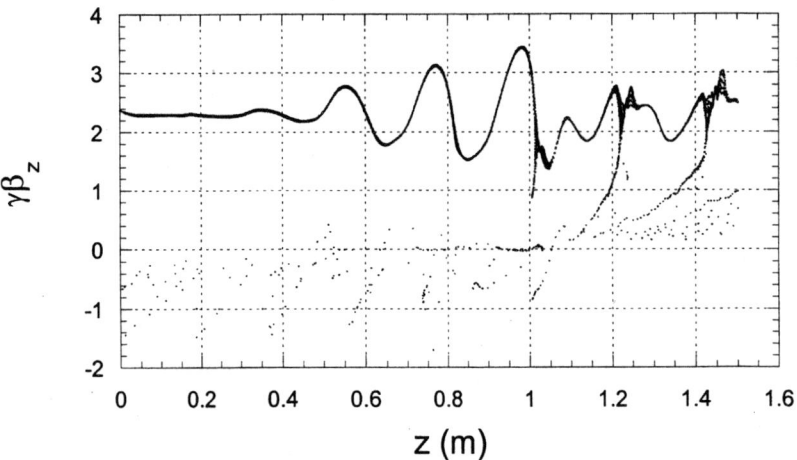

FIGURE 5. Beam energy versus axial position z along the ABK with a capacitively tuned (-220 MHz) penultimate cavity. Output cavity location is at z=1.2 m. Note the much smaller energy spread at z = 1.2m.

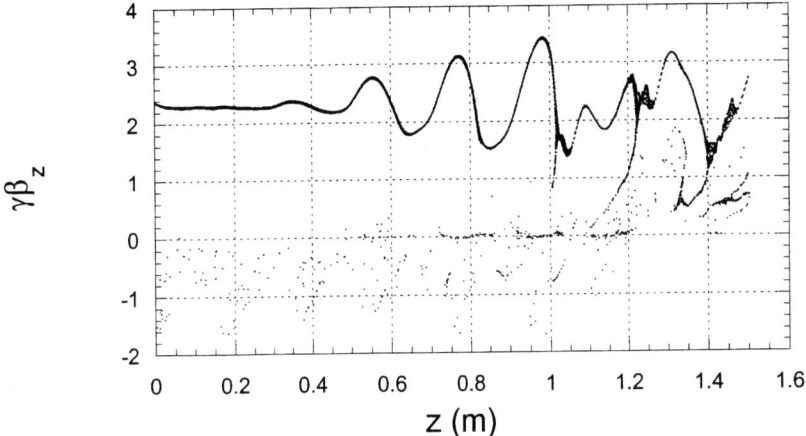

FIGURE 6. The longitudinal phase space of the beam for the high-efficiency narrow gap output cavity case. Even with the high output cavity voltage, a virtual cathode is not seen.

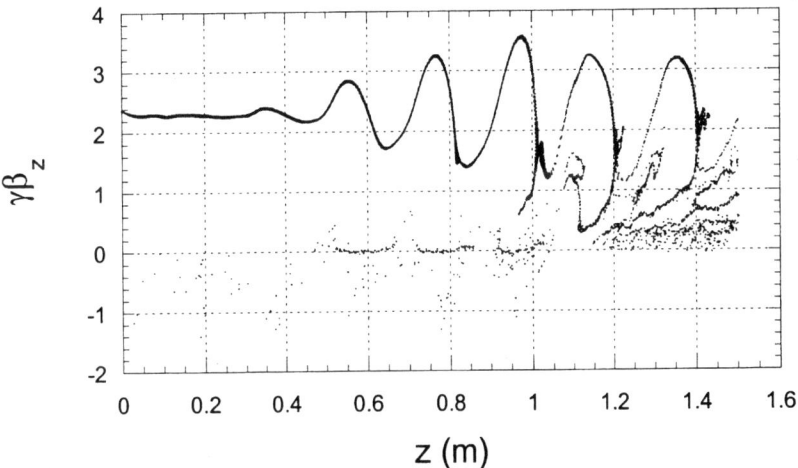

FIGURE 7. The longitudinal phase space equivalent to Fig. 6 except that the penultimate cavity is inductively tuned instead of capacitively tuned. The large energy spread limits the extraction efficiency to about 23%, larger body current is generated, and a large slug of charge forms a deleterious virtual cathode between the output cavity and the collector.

in the beam's energy spread, which translates to being able to run a larger output cavity voltage without stopping or turning around the lower energy electrons. Even without an output cavity, a virtual cathode formation is evident in Fig. 4 at 1.2 m, which is absent in Fig. 5.

The first output cavity installed on the ABK will be designed to minimize the risk of RF breakdown by using an extra wide gap output cavity that couples RF from the beam with low efficiency. The induced current will be reduced to about 35% from the increase in transit time across the output gap, the output cavity voltage will be 500 kV, and the extraction efficiency will be only 15%. Although this approach results in a lower output power, we are more likely to obtain a better understanding of the relevant physics with the reduced risk of RF breakdown. For reaching full power in the second iteration, a "high-performance" narrow gap output cavity will be installed. The increase in transit time factor will lead to a 52% induced current and 750 kV in the output cavity, and an extraction efficiency of about 30%. The decrease in induced current from 56% to 52% is due to the presence of the output cavity fields. The longitudinal phase space of the beam for the high-efficiency narrow gap case is shown in Figure 6. Even with the high output cavity voltage, a virtual cathode is not seen in Fig. 6. For comparison, Fig. 7 shows the longitudinal phase space equivalent to Fig. 6 except that the penultimate cavity is inductively tuned instead of capacitively tuned. The large energy spread limits the extraction efficiency to about 23%, larger body current is generated, and a large slug of charge forms a deleterious virtual cathode between the output cavity and the collector, which is already partly present in Fig. 4, but essentially absent from Fig. 6. The use of capacitive penultimate cavity tuning will be important to ABK efficiency.

ABK OUTPUT CAVITY DESIGN CONSIDERATIONS

The output cavity produces microwave energy by converting the kinetic energy of the modulated electron beam. The cavity shunt impedance (R/Q) is important because the cavity voltage must be high enough to extract all available kinetic energy of the beam but not so high that electrons are reflected, leading to oscillations. The output cavity design represents a series of engineering tradeoffs. Symmetry of the field pattern in the region of the beam is also important. Electrons around the circumference of the annular beam must be acted on equally by the decelerating electric field, which is especially critical for annular beam devices because of the beam's proximity to the wall. Multiple outputs must be used because of the high power levels, and each output affects the decelerating field profile to some degree. Surface electric fields in the cavity must be kept as low as possible for the given field strength. Surfaces with high electric fields must be adequately radiused. Re-entrant cavity designs are not typically used for high-power output cavities. Instead, noseless gap designs are used. The output cavity and its output waveguides must fit within the available space on the magnets and in the klystron oven for vacuum baking.

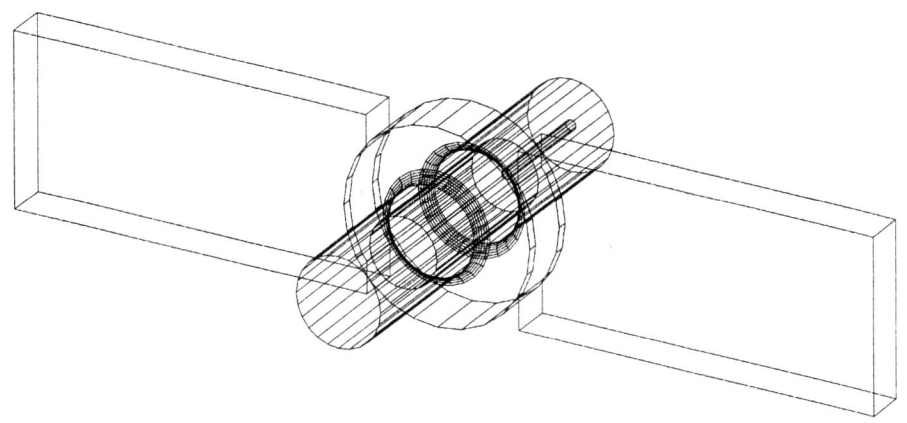

FIGURE 8. ABK output cavity with two waveguide output arms.

Full 3-D electromagnetic modeling was necessary to address the heavily loaded output coupling. A wide gap was chosen to minimize the chance for arcing. A wide gap will not extract the beam's energy as efficiently, but a wide gap will have much lower electric field stresses and be less likely to suffer RF breakdown. The gap cannot be arbitrarily widened, however, as this would increase the probability for forming a virtual cathode in the output cavity gap because the beam potential depression increases in the gap due to the increased wall radius. The output cavity is 6 cm wide with two 4 cm tall, low-impedance, WR-650 waveguide arms to couple the RF power out as shown in Fig. 8. We are aiming for a shunt impedance of about 300-400 Ω and a Q of less than 5. This should give the correct voltage to decelerate the beam bunches without returning electrons and have surface fields low enough (< 300 kV/cm) to avoid breakdown across the gap. For the cavity shown, and a modulated current I_1 of 2 kA, the beam deceleration voltage would be 862 kV. Assuming the transit time factor is about 60%, this would be effectively about 500 kV. The power out each waveguide arm would be about 400 MW, and the peak field stress would be less than 132 kV/cm.

SUMMARY

A beamstick for the annular beam klystron has been built comprising a temperature-limited thermionic electron gun, two modulating cavities, and the beam collector. The beam stick has not been successfully tested yet due to a heater winding failure; however the thermal characteristics of the cathode and heater assembly have been measured and meet the design requirements. A low voltage perveance test was successful indicating a reasonable gun design. The RF design of the complete tube is essentially finished and indicates that strong capacitive detuning of the penultimate cavity is necessary to obtain a low energy spread and reasonable efficiency. We are currently preparing to rebuild the heater/cathode assembly and fabricate the complete RF circuit.

ACKNOWLEDGMENTS

We would like to thank Dr. George Caryotakis at SLAC for making his personnel and facilities at SLAC available for the design and fabrication of the annular beam klystron. The authors also want to express their sincere gratitude to the Los Alamos technicians, Aaron Archuleta, Joe Strotman, Houston Martinez, and Steve Klein for their hard work in setting up the modulator test facility and for their unflagging enthusiasm throughout the testing effort. Work supported by the Los Alamos National Laboratory Directed Research and Development Program

Relativistic Magnetron With Diffraction Antenna

Mikhail I. Fuks and Edl Schamiloglu

Department of Electrical & Computer Engineering
University of New Mexico, Albuquerque, NM 87131, USA

Abstract. The relativistic magnetron with diffraction output is one of the few compact high power microwave sources that is compatible with low-impedance pulsers. The simplest design of such a source is a magnetron resonant system that is gradually tapered outward to a conical horn antenna. In this present work it is shown that the use of a diffraction output allows one to minimize the magnetic field-producing system, provide for selection of longitudinal modes, increase the interaction space, and accordingly the operating current and output power.

INTRODUCTION

The relativistic magnetron oscillator is a powerful, reliable, and compact source of coherent radiation in the centimeter and decimeter wavelengths. Very large radiated powers of $P \approx 10$ GW in L-band [1] and $P \approx 4$ GW in X-band [2] have been generated in interaction spaces with volume $V < \lambda^3$ (where λ is the operating wavelength) in experiments reported about twenty years ago. However, in spite of such attractive parameters, the relativistic magnetron has attracted little subsequent interest. Some reasons for this are described below.

It is well-known in the traditional vacuum electronics community that the lifetime of magnetrons is typically shorter by an order of magnitude compared with the lifetime of devices driven by transient electron beams. The reason for the shorter magnetron lifetime is electron bombardment of the electrodes directly inside the interaction space. For relativistic magnetrons powered by a unipolar pulse generated from a pulser through a coaxial line (see (5) in Fig. 1) the problem of lifetime is dramatized by the inhomogeneous distribution of electron bombardment on the surface of electrodes. This is caused by axial electron drift in a crossed radial electric field and the axial flow's azimuthal self-magnetic field. Experimental X-band relativistic magnetrons [2, 3] failed to operate after a few hundred shots because of the erosion of vanes in the resonator systems.

The traditional design of a relativistic magnetron, as for example the famous A6 magnetron [4], comprises a system of two coaxial electrodes immersed in an axial magnetic field. One electrode is a cold explosive emission cathode (1) and the other electrode is an anode block (2) with resonators (Fig. 1). Radiation is extracted using waveguide (3) from one of the resonators through a narrow coupling slot (4). Such a

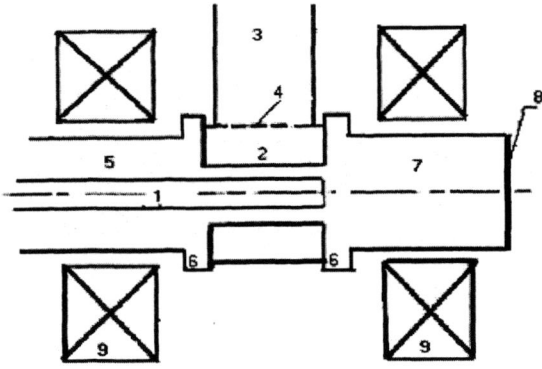

FIGURE 1. Traditional design of a relativistic magnetron: 1 - cathode; 2 - resonant system; 3 - output waveguide; 4 - coupling slot; 5 - coaxial line; 6 - side cavities; 7 - additional volume; 8 - electron dump; 9 - pair of Helmholtz coils.

simple design for a source cannot seemingly be an obstacle for applications, but we shall see what the inherent limitations of this source are.

As a relativistic high power microwave generator the relativistic magnetron's primary advantage is output power. Although the relativistic magnetron has achieved 10 GW power levels, the possibility for increasing this further is strongly limited in a traditional design.

In order to increase the radiated power, one must increase the volume of the interaction space (that is, increase the number of resonators). However, in this case fast condensation of eigenmodes can interrupt single mode operation. The problem of mode selection is further complicated by the non-symmetrical radiation output; reflections change the phase relations between resonators that can lead to mode hopping from π-type oscillations to a neighboring mode. All of the eigenmodes, excluding π-type oscillations, are degenerate with sinusoidal and cosinusoidal azimuthal distributions, but the non-symmetrical output removes the degeneracy, fixing the azimuthal position of nodes and antinodes. This results in a lower quality of the radiated wave having an antinode at the coupling slot compared with the quality of a locked-up mode having a node at the coupling slot. This, in turn, leads to the splitting of the spectrum of these waves and the excitation of the locked-up one that can result in pulse shortening, and the subsequent overheating and destruction of the magnetron. The danger of mode hopping does not permit one to increase the number of resonators up to natural limit, which is determined by overlapping of bands of synchronous interaction of neighboring eigenmodes.

In order to facilitate higher current operation it is necessary to increase both the cross section and length of the interaction space. However, the axial electrical length is limited by the value $L \approx (0.5 - 0.7)\lambda$ in order to avoid competition with longitudinal waves. (By the way, owing to its small axial length and referring to Fig. 1, additional side cavities (6) are used for equalization of the axial electric field.) As a result of the limitation of the inter-electrode volume, the impedance the electron beams used does not differ from linear devices driven by transient beams, which usually have an impedance $Z \geq 100\,\Omega$. This means that the ability to operate a traditional magnetron using higher current (lower impedance) pulsers is limited.

Oqwing to the asymmetric output of the source (3) (the numbers in parenthesis all refer to Fig. 1) a pair of Helmholtz coils (9) is used to create the axial magnetic field H_z required for the azimuthal electron drift that is synchronous with the rotating wave. Application of such a large magnetic system to provide the magnetic field inside a small interaction space between electrodes (1, 2) offsets the magnetron's compactness.

Additional volume (7) between the resonant system (2) and the electron dump (8) is included in order to decrease the axial electron current since the space-charge-limiting current will decrease with increasing volume. At the same time this cavity must be decoupled with the electromagnetic fields of the resonant system.

Finally, in a traditional magnetron the output power will ultimately be limited by microwave breakdown in the narrow coupling slot (4).

MAGNETRON WITH DIFFRACTION OUTPUT

One way of realizing a compact high power relativistic magnetron is to use a diffraction output. Although the relativistic magnetron with diffraction output was suggested and realized in the 1970s [3], at that time it was not perceived as the most natural realization of a relativistic magnetron [5]. Here we try to describe some unique possibilities for such magnetrons that not only do not suffer the drawbacks of traditional relativistic magnetrons, but also have many other advantages that make these magnetrons attractive for high power applications.

The simplest design of a relativistic magnetron with diffraction output is shown in Fig. 2, where an actual photograph of a 4 GW X-band magnetron is presented [2]. Referring to this figure, the resonator system (2) is the symmetrical load of the coaxial line (4) that is used to power the magnetron. All resonators of the anode block are tapered outward onto a conical horn antenna (3). In this case the radiation pattern is determined by the structure of the operating mode of the magnetron (Fig. 3a).

When the operating mode is π-type, the radiated wave is TE_{nl} where $n = N/2$, N is the number of resonators. In order to radiate, all resonators on the conical surface must be continued up to the radius of the horn antenna, which exceeds the radius corresponding to the cutoff frequency of the radiated wave. To minimize reflections from the horn antenna, its opening angle α must be small ($\alpha \ll \lambda/D$ where D is the diameter where the magnetron joins the antenna [6]).

Such a diffraction output, similar to the ones used in powerful gyrotrons, provides a high resistance to breakdown and a low quality factor Q, which is close to the minimal diffraction Q_{dif}-factor [7]

$$Q \approx Q_{dif} = \frac{8\pi}{m}\left(\frac{L}{\lambda}\right)^2.$$

Therefore, the maximal Q-factor is realized for the simplest axial structure of the electromagnetic field ($m = 1$) that automatically solves the problem of selection of longitudinal modes independent of the axial electric length L/λ of the magnetron. Unlike the traditional magnetron, for the magnetron with diffraction output there is the possibility of selecting the desired amplitude $\tilde{E}$ of the operating mode, and by doing so, the anode current I_a and the output power P as well, simply by selecting the appropriate axial length L, as it is shown from energy balance considerations $\omega W/Q = P$, taking into account that the energy W accumulated in the resonant system is proportional to $W \sim \tilde{E}^2$ and the anode current is proportional to $\tilde{E}$ through $I_a = A\tilde{E}$ (the coefficient A is determined by conditions related to the formation of electron spokes through the mechanism of electron emission [8]). Here $P = \eta U I_a$, where U is the anode voltage and η is the magnetron efficiency given by $\eta = 1 - 2R_H/d = 1 - \left(2mc^2/eU\right)\beta_{ph}^2/\left(1 - \beta_{ph}^2\right)$ [8], which is determined by the electron Larmor radius R_H on the cathode surface and the inter-electrode gap d or phase velocity v_{ph} of a synchronous wave $\beta_{ph} = v_{ph}/c$. Finally, c is the speed of light, e and m are the electron charge and rest mass, respectively, and ω is the angular frequency.

The fact that each cavity provides an equal load in this design allows for the possibility of any mode becoming the operating mode (for some applications the use

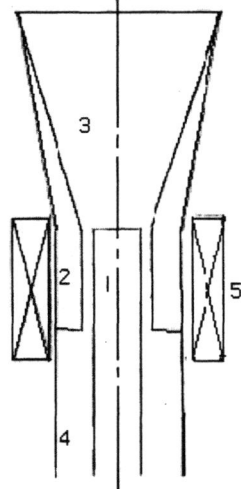

Fig. 2. Magnetron with diffraction radiation output: 1 - cathode; 2 - resonant system; 3 - horn antenna; 4 - coaxial line; 5 - solenoid.

of a rotating wave, providing an azimuthally homogeneous radiation pattern, can be advantageous). Furthermore, it is possible to increase the number of resonators up to the maximum allowable number (that is, up to the overlapping of bands of synchronous interaction), which is determined by the lifetime of electrons in the inter-electrode gap.

Increase of the interaction volume (both the cross section and the length) gives the possibility to increase the output power by using higher current (in experiment [2] the electron beam with the impedance $Z \approx 25\,\Omega$ was used without any optimization of magnetron parameters).

An axially symmetric radiation output allows one to minimize the magnetic field-producing system. A pulsed solenoid around the resonator system (Fig. 2) alone is sufficient in this case, and leads to a compact source of high power microwaves.

The maximum diameter of the magnetron is determined by the diameter of the output waveguide. This can be reduced by converting the output wave directly within the horn antenna. The principle of such a mode conversion can be demonstrated when considering the example of a magnetron with N = 12 resonators (Fig. 3). Let us consider π-mode operation. When only alternate resonators are continued on the conical surface, cophased oscillations will be excited. The resultant radiation is the symmetrical TE_{01} wave. Accordingly, the diameter of the output waveguide can be decreased by a factor of 2.5. There are other possibilities of transforming the radiation pattern. When only every third resonator is continued onto the conical surface, the structure of the output radiation will correspond to the TE_{21} wave.

Note that the electron current that is leaving the interaction space in the axial direction (a current that is useless for magnetron operation) is naturally limited by its own space charge in the radiated antenna. This is unlike the situation in traditional magnetrons, where a special cavity is required for this purpose.

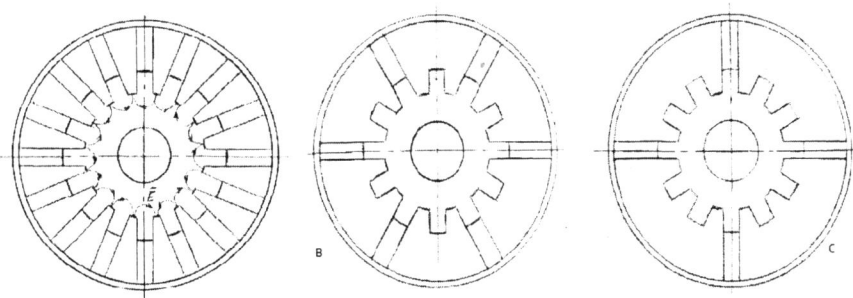

Fig. 3. Mode converters on the horn antenna for a 12-resonator magnetron with π-mode operation. Patterns for antennas (A), (B) and (C) correspond to the TE_{61}, TE_{01}, and TE_{21} radiated waves, respectively.

CONCLUSION

The relativistic magnetron with diffraction radiation output with evenly loaded resonators is the most natural design for relativistic electron energies because:

1. it is highly resistant to microwave (electrical) breakdown;
2. there is an absence of dangerous mode skipping to unloaded-type oscillations (this allows one to increase the number of resonators up until an overlap of the synchronism bands, and to use any oscillation as the operating mode);
3. its operation is based on the simplest axial distribution of the electric field independent of the axial length of the magnetron; this allows one to choose the length of the device based only on the required amplitude of the microwave field;
4. the (useless) axial current is naturally limited by its own space charge in the horn antenna;
5. the volume of magnetic field required is minimal compared to a traditional magnetron;
6. one can transform the output radiation pattern directly onto the diffraction antenna.

The relativistic magnetron with diffraction antenna is an attractive load for an explosively-driven generator.

ACKNOWLEDGEMENTS

This work was supported by an AFOSR/ New World Vistas Grant.

REFERENCES

1. Didenko, A.N., Fomenko, G.P., Gleizer, I.Z.,Krasik, Ya.E., Mel`nikov, C.V., Perelygin, S.P., Shtein, Yu.G., Sulakshin, A.S., Tsvetkov, V.I., and Zherlitsin, A.G., "Generation of high power RF-pulses in magnetron and reflex triode systems," in Proceedings of the 3rd Int. Topical Conference on High Power Electron and Ion Beam Research and Technology, Novosibirsk, Soviet Union, 1979, vol.2, pp. 683-691.
2. Kovalev, N.F., Krastelev, E.G., Kuznetsov, M.I., Maine, A.M., Ofitserov, M.M., Papadichev, V.A., Fuks, M.I., and Chekanova, L.N., "High power relativistic 3 cm magnetron," Sov. Tech. Phys. Lett., 6, 1980, pp. 4-8.
3. Kovalev, N.F., Kol`chugin, B.D., Nechaev, V.E., Ofitserov, M.M., Soluyanov, E.I., and Fuks, M.I. , "Relativistic magnetron with diffraction coupling," Pis`ma v ZhTF, 3, 1977, pp. 1048-1051.
4. Bekefi, G., and Orzhechowski, J.J., "Giant microwave burst from a field-emission, relativistic-electron-beam magnetron," Phys. Rev. Lett., 37, 1976, pp. 379-382.
5. Fuks, M.I., and Kovalev, N.F., "Relativistic magnetron with diffraction output," Program and abstracts of 11th Int. Conf. on High-Power Electromagnetics, Tel-Aviv, Israel, June 14-19, 1998, p. 18.
6. Katsenelenbaum, B.Z., "Theory of irregular waveguides with smoothly changing parameters," Moscow, AS USSR, 1961, p. 216.

7. Vlasov, S.N., Zhislin, G.M., Orlova, I.M., and Petelin, M.I., "Open resonators in the form of waveguides with variable sections," Izv.VUZov, Radiofizika, 12, 1969, pp. 1236-1244.
8. Nechaev, V.E., Petelin, M.I., and Fuks, M.I., "Prospects for magnetron devices with relativistic electrons," Pis`ma v ZhTF, 3, 1977, pp. 763-767.

Review of Computational Models for High Power Microwave Sources

Eric M. Nelson

MS B259, Los Alamos National Laboratory, Los Alamos, NM 87545

Abstract. Numerous computer models have been developed for the analysis and design of high power microwave sources. The models include electrostatic gun codes, beam-circuit interaction codes, and pure electromagnetic codes. The beam-circuit interaction codes include steady-state and time-domain models. They include particle-in-cell and gyrokinetic codes. The electromagnetic codes include eigenmode, frequency-domain and time-domain solvers. Fields are computed using the finite-difference and finite-element methods. The state of the art for such models is reviewed, and suggestions for future work are presented.

INTRODUCTION

Numerous computer models have been developed for the analysis and design of high power microwave (HPM) sources. First, I will describe the most common types of models. For each type of model I will compare and contrast various codes that implement the model. Second, I will describe relationships between the models and comment on how they are used in practice. Finally, I will discuss the goals of an HPM computational program and suggest directions for future work.

MODELS AND CODES

Numerous models are employed to analyze and design HPM sources. In each subsection below, I briefly describe a model and provide some examples of codes that implement the model.

The first three models below are particle-in-cell (PIC) codes, large-signal codes and gun codes. These three models self-consistently compute the electromagnetic fields and the trajectories of the beam's charged particles. PIC codes make few assumptions about the electromagnetic field or the electron beam. Large-signal codes typically restrict the electromagnetic field to select frequencies, and gun codes further restrict the electromagnetic field to be static.

The next two models are electromagnetic codes and magnetostatic codes. The electromagnetic codes include eigenmode solvers, frequency-domain solvers and time-domain solvers. These models compute electromagnetic fields from specified sources, which might include currents, charges and/or fields at waveguide ports.

The final model below is a special-purpose model for multipactor studies. In this model, charged particles are tracked in specified electromagnetic fields.

Particle-in-Cell Codes

The electromagnetic particle-in-cell (PIC) code is the most general model applied to HPM sources. They include features and behavior missing from large-signal codes and gun codes. For example, PIC codes can reproduce instabilities that would be missed by the other two models. PIC codes typically handle complicated geometries and they often contain a variety of specialized physics packages, such as secondary emission and ionization of a background gas.

When computing steady-state solutions, PIC codes are slower than large-signal codes. When computing static solutions, PIC codes are slower than gun codes.

A typical electromagnetic particle-in-cell (PIC) code is identified by three features: (1) a finite-difference time-domain (FDTD) electromagnetic field solver, (2) macroparticles for representing the beam, and (3) the fields push the macroparticles and the macroparticles drive the fields. The classic references on PIC codes are [1, 2].

Numerous PIC codes have been applied to HPM source analysis and design. Some codes and a few of their applications are listed below.

- XOOPIC [3, 4], from University of California (UC) at Berkeley, is an object-oriented parallel PIC code with collision and secondary models. It has been used to study pulse shortening in a relativistic klystron oscillator (RKO). Work is in progress on a 3D version.
- MAGIC [5], from Mission Research Corporation (MRC) [6], is a commercial $2\frac{1}{2}$D and 3D PIC code with numerous physics modules. The $2\frac{1}{2}$D version was recently used to improve the cathode design of a magnetically insulated line oscillator (MILO) [7].
- ICEPIC [8], from the Air Force Research Laboratory (AFRL), is a parallel 3D PIC code with recent enhancements [9] to their space-charge-limited emission algorithm.
- EMX [10], from Culham [11], is a multiblock body-fitted parallel 3D PIC code. It was recently used to optimize a MILO [12] and model relativistic magnetrons.
- TWOQUICK and QUICKSILVER, from Sandia National Laboratory (SNL), are $2\frac{1}{2}$D and 3D PIC codes, respectively. The codes have been used to model MILOs. The 3D code has been enhanced recently with perfectly matched layer (PML) boundary conditions [13].
- MAFIA, from Computer Simulation Technology (CST) [14], is a commercial software package that includes a 2D and a 3D PIC code. The 3D PIC code was employed in a recent study of asymmetries in a traveling wave tube (TWT) with periodic permanent magnet (PPM) focusing [15].
- MASK and ARGUS, from SAIC [16], are $2\frac{1}{2}$D and 3D PIC codes, respectively. MASK and the closely related code CONDOR continue to be applied to high power klystrons.
- ISIS, from LANL, is a body-fitted $2\frac{1}{2}$D PIC code with recent applications to an annular beam klystron (ABK). There was also a parallel 3D version at one time.

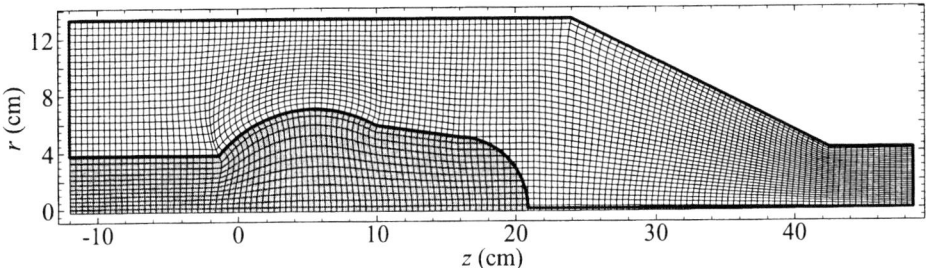

FIGURE 1. An example of a body-fitted grid from ISIS. This is a magnetron injection gun. The shaded area is conducting. The shallow slope of the cathode (at $10\,\text{cm} \leq z \leq 17\,\text{cm}$) is well resolved without stairstepping and with only a modest size grid.

Most PIC codes employ a cartesian grid that leads to a stair-stepped representation of curved and sloped surfaces. Two of the PIC codes, EMX and ISIS, employ a body-fitted grid to improve the representation of such surfaces and thus improve the accuracy of the solution. An example of a body-fitted grid is shown in figure 1. The shallow slope of the magnetron injection gun's cathode is well resolved.

Large-Signal Codes

Like a PIC code, a large-signal code pushes macroparticles with electromagnetic fields and drives electromagnetic fields with macroparticles. However, a large-signal code differs from a PIC code in three ways: (1) the electromagnetic field is restricted to a linear superposition of a modest number of modes, (2) the device configuration and geometry is restricted to a specific class, and (3) only steady-state solutions are computed.

A large-signal code computes a steady-state solution for an HPM source faster than a PIC code, but it does so with some simplifications. The space-charge fields are often less accurate in the vicinity of the output structure, and reflected electrons are handled with difficulty. A large-signal code is slower than a small-signal (semi-analytic) code, but the large-signal code includes more nonlinear effects which are missing from small-signal codes.

Numerous large-signal codes have been developed and applied to HPM source analysis and design. Some of the codes and a few of their applications are listed below.

- MAGY [17] and MAGYKL [18], from the University of Maryland (UMD) and the Naval Research Laboratory (NRL), are large-signal codes for gyrotrons and gyroklystrons, respectively. They solve a multimode telegrapher's equation for the electromagnetic fields. Highly lossy walls were recently added to MAGY [19].
- CHRISTINE [20], from UMD, NRL and SAIC, is a large-signal code for helix TWTs. Unlike most large-signal codes, CHRISTINE is polychromatic; it handles signals at multiple non-harmonic frequencies simultaneously. It is also an extensively validated code. A 3D version is under development.

- GATOR [21], from SAIC, is a large-signal code for helix and coupled-cavity TWTs which employs a PIC model to compute space-charge fields.
- ISIS, a PIC code from LANL, has been converted to a large-signal code for klystrons and TWTs.
- CONDOR/MASK, a PIC code from SAIC [16], is often employed as a large-signal code for klystrons.

There are many other large-signal codes which I will not attempt to list. These are often based on simple disk models (e.g., JPNDISK) and many of them are not widely available because they are proprietary.

Gun Codes

A gun code is used for the design and analysis of electron guns. Gun codes are also used to model collectors. They self-consistently solve for the static electric and magnetic fields generated by a beam. They are similar to PIC codes because the fields push the macroparticles and the macroparticles influence the fields. Gun codes also handle complicated geometries. But, unlike PIC codes, the fields are static; there is no time dependence.

Both finite difference and finite element methods are employed for the electrostatic field solution. The finite difference method is fast and easy to implement, but the finite element method can provide better accuracy. Algorithms for computing the self-magnetic field vary. Most gun codes compute only the azimuthal self-magnetic field. A few gun codes are listed below.

A gun code arrives at a static field solution and beam parameters faster than a PIC code. But a PIC code might still be employed to verify that an electron gun design is stable and correct.

- EGUN [22], from SLAC, is the classic electron gun code. It is a 2D finite difference gun code. UGUN, from Raytheon, is similar. EGN2e is a commercial version of EGUN.
- DEMEOS [23], from Litton, and TRAK, from Field Precision [24], are 2D body-fitted finite element gun codes. These codes employ a conformal structured triangular mesh.
- MICHELLE, from SAIC and NRL, is a 3D finite element gun code. It employs both multiblock body-fitted structured meshes and general unstructured meshes. It includes extensive secondary emission models for modeling collectors. A novel particle tracker for unstructured grids has recently been deployed, and a 2D version is under development. An example of a gridded gun simulation is shown in figure 2.
- SCALA, from Vector Fields [25], is a commercial 3D finite element gun code which employs both structured and unstructured meshes.
- ARGUS and AVGUN, from SAIC [16], are 3D finite difference gun codes. MAFIA, from CST [14], also contains a 3D finite difference gun code.

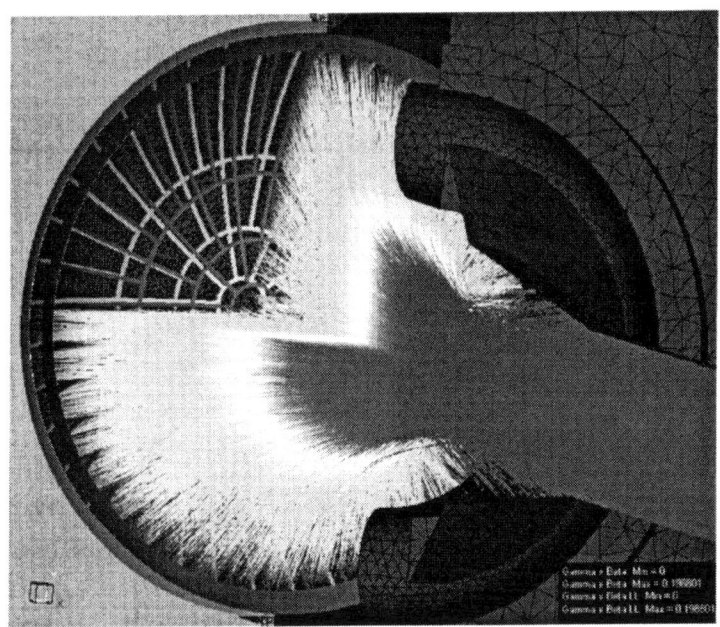

FIGURE 2. An example of a gridded gun simulation with the 3D finite element gun code MICHELLE. A portion of the beam has been cut away to reveal the control grid and the shadow grid.

- COCA [26], from the University of Catania, is a 3D finite element gun code with adaptive mesh refinement. It has been applied to collector design.
- SAPPHIRE, from Culham [11], is a 3D finite element gun code which employs body-fitted multiblock structured meshes. It includes collision models for ions.

Electromagnetic Codes

Electromagnetic codes solve Maxwell's equations in complicated geometries with prescribed sources, if any. There are three basic types of electromagnetic codes: eigenmode solvers, frequency-domain solvers and time-domain solvers.

Eigenmode solvers are typically used for cavity design. They compute the frequencies and field patterns of electromagnetic modes. There is no source current or charge. Both finite element and finite difference methods are employed. Very high accuracy can be achieved.

Frequency-domain solvers are typically used for waveguide components, input cavities and output structures. They compute the harmonic fields at a specified frequency given source currents, charges, and/or waveguide port excitations. Both finite element and finite difference methods are employed.

Finite difference time-domain solvers are often employed as an alternative to frequency-domain codes.

- SUPERFISH, available from the Los Alamos Accelerator Code Group (LAACG) [27], is a classic 2D finite element eigenmode solver which employs a conformal structured triangular grid.
- MAFIA, from CST [14], contains 3D finite difference time-domain, frequency-domain and eigenmode solvers.
- GdfidL, from TU-Berlin, is also a 3D finite difference time-domain and eigenmode solver.
- HFSS, from Ansoft [28], is a 3D finite element frequency-domain and eigenmode solver with adaptive mesh refinement and a variety of constitutive properties.
- CTLSS [29], from SAIC, is a 3D body-fitted multiblock frequency-domain and eigenmode solver with a robust solver that handles lossy materials well.
- YAP, developed at SLAC and LANL, is a 3D finite element eigenmode solver which employs high order elements to achieve very high accuracy. This was further developed at SLAC to produce OMEGA. It includes adaptive mesh refinement and a parallel solver. An improved interface was developed by STAAR [30], and it is currently available from them as the code ANALYST.

There are many other electromagnetic codes, especially those in the antenna and radar cross section (RCS) communities. There are also numerous scattering matrix and mode-matching codes. These codes trade geometric flexibility for speed. CASCADE is one example.

Magnetostatic Codes

Magnetostatic codes solve for the static magnetic fields in complicated geometries given source currents and/or permanent magnets. They are typically used to compute the guiding field for beam transport. They often include nonlinear saturation effects in magnetic materials.

- Poisson, from LAACG [27], is the classic 2D finite element magnetics code. It employs a conformal structured triangular grid. TRAK, from Field Precision [24], employs similar algorithms.
- EMAG, from ANSYS [31], and MAXWELL, from Ansoft [28], are commercial 3D adaptive finite element magnetostatic codes.
- mPPM and lesPPM, from SAIC [16], are examples of special-purpose magneto-statics codes. It is a semi-analytic model for periodic permanent magnet (PPM) systems. End effects are included.

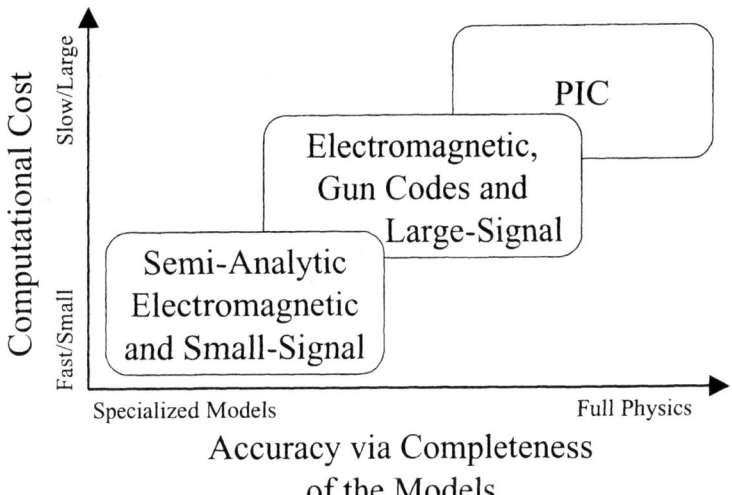

FIGURE 3. A hierarchy of HPM computational models. The horizontal axis is accuracy due to the completeness of the physical models. The vertical axis is computational cost per run.

Multipactor Codes

Multipactor codes track particles with secondary emission in the presence of prescribed fields. There are various codes for special geometries, such as coaxial lines. Some PIC codes have been applied to multipactor simulations in simple geometries. They can include the growth of space charge in the model. Two codes, TRAK from Field Precision [24] and MUPAC [32] from CEA, model multipactor in arbitrary 2D geometries.

CODE RELATIONSHIPS AND USE

Figure 3 shows an approximate hierarchy of models employed in the design and analysis of HPM sources. The horizontal axis represents the accuracy of a model due to the completeness of the physics it reproduces. The vertical axis represents the computational cost per run of a model. The computational cost is not just CPU time, memory and disk space. It is also user effort for training, problem setup (including geometry specification and mesh generation) and postprocessing (analysis of the model output). It could also include the code development effort, which would be amortized over the use of the code.

The semi-analytic electromagnetic and small-signal codes are highly specialized, and they tend to be fast and small. The electromagnetic, large-signal and gun codes are moderately specialized models that typically require modest computational resources. The PIC codes are the most general model, and they typically consume the most computational resources.

Figure 4 shows a hierarchy of use for these models. The vertical axis now represents

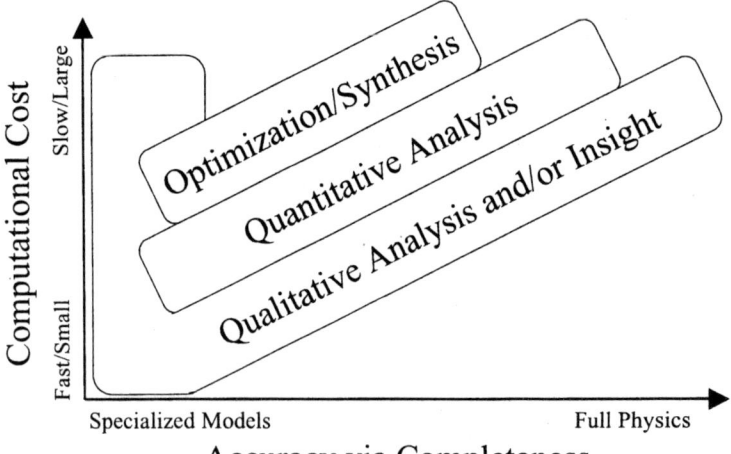

FIGURE 4. A hierarchy of use for HPM computational models. The horizontal axis is accuracy due to the completeness of the physical models. The vertical axis is computational cost for a design.

the computational cost per design. The most specialized models typically provide only qualitative analysis and/or insight because they lack the physics and/or geometry needed to produce measurably accurate results in real HPM sources. The electromagnetic, large-signal and gun codes do provide measurably accurate results with a moderate amount of work. They are also used extensively to optimize the components of an HPM source. PIC codes are computationally expensive, so they are typically used at most to quantitatively verify a specific design, and perhaps to optimize the parameters for output structures. The complete design of an HPM source is rarely optimized with a PIC code. On the other hand, they are often employed for qualitative analysis and/or insight.

Consider the design of a klystron circuit as an example. One first chooses approximate cavity parameters using analytic formulas and perhaps a small-signal code. The design of the klystron circuit is then optimized using large-signal codes. The physical geometry of the output structure and the cavities are designed using electromagnetic codes. Finally, the design is verified using a PIC code and perhaps a small-signal dipole mode calculation. The work flow is typically iterative. Constraints from later portions of the design cycle (e.g., cavity design with electromagnetic codes) are fed back to earlier portions (e.g., optimization of the circuit with large-signal codes).

GOALS AND SUGGESTIONS

The motivation for an HPM computational program is to develop a better HPM source faster and cheaper. This can be accomplished in two ways.

First, by faithfully, easily and rapidly modeling known physics, one can focus the development effort on those problems which do require experimentation and trial and error. One such problem is pulse shortening. A better computational program would allow for more effort to be placed on the diagnosis of pulse shortening mechanisms, for example. There are also numerous engineering and technology issues involving materials, cost and system design which benefit from increased attention.

Second, by providing cheap, effective tests of hypotheses, one can improve the design of those experiments which are required for the development of an HPM source.

There is much room for improvement in the computational models and their implementations. In general, the models are improved by reducing their computational cost (computer time and memory but also user effort) and/or expanding their capabilities (not only physics but also numerical accuracy). Here are some suggestions.

For PIC codes: develop better load balancing for particle pushing in parallel codes; extend the use of body-fitted multiblock codes; employ improved software engineering practices, including documentation; and develop a testbed paradigm where one can easily add and change components.

For gun codes, add 3D self-magnetic fields for multibeam klystrons (MBKs) and sheet beam klystrons (SBKs).

Ideally, large-signal codes should do most of the design work, leaving PIC codes for design verification. For klystrons: improve the treatment of space-charge fields in output structures (a hard problem); employ improved software engineering to facilitate connections with other codes, such as optimizers; and adapt the codes for MBKs and SBKs.

Perhaps most importantly, those physicists and electrical engineers that develop codes should be better trained in software engineering. There is much experience in other communities, and we should take advantage of it. This can be informal, such as by following some good examples. We can provide suggested reading lists. We can emulate successful development and maintenance programs, especially for codes which were developed intermittently and in a distributed fashion, which is a common situation for code development in the research community.

I will conclude with the following remarks. Computation with existing models is an essential component of HPM research and development programs. However, there is still a lot of room for improvement of the computational models, and we could be doing a better job developing improved computational models.

ACKNOWLEDGMENTS

This work was supported by DOE, contract W-7405-ENG-36, and ONR, NRL contracts N00014-97-C-2076 and N00173-98-MP-00127.

REFERENCES

1. Hockney, R. W., and Eastwood, J. W., *Computer Simulation Using Particles*, IOP Publishing Ltd, Bristol and Philadelphia, 1988.
2. Birdsall, C. K., and Langdon, A. B., *Plasma Physics via Computer Simulation*, Institute of Physics Publishing, Bristol and Philadelphia, 1991.

3. Verboncoeur, J. P., Langdon, A. B., and Gladd, N. T., *Computer Physics Communications*, **87**, 199–211 (1995).
4. Visit http://ptsg.eecs.berkeley.edu/.
5. Goplen, B., Ludeking, L., Smithe, D., and Warren, G., *Computer Physics Communications*, **87**, 54–86 (1995).
6. Visit http://www.mrcwdc.com/magic.
7. Haworth, M. D., Luginsland, J. W., and Lemke, R. W., *IEEE Transactions on Plasma Science*, **29**, 388–392 (2001).
8. Blahovec, J. D., Jr., Bowers, L. A., Luginsland, J. W., Sasser, G. E., and Watrous, J. J., *IEEE Transactions on Plasma Science*, **28**, 821–829 (2000).
9. Watrous, J. J., Luginsland, J. W., and Sasser III, G. E., *Physics of Plasmas*, **8**, 289–296 (2001).
10. Eastwood, J. W., Arter, W., Brealey, N. J., and Hockney, R. W., *Computer Physics Communications*, **87**, 155–178 (1995).
11. Visit http://www.culham.com/em/.
12. Eastwood, J. W., Hawkins, K. C., and Hook, M. P., *IEEE Transactions on Plasma Science*, **26**, 698–713 (1998).
13. Pasik, M. F., Seidel, D. B., and Lemke, R. W., *Journal of Computational Physics*, **148**, 125–132 (1999).
14. Visit http://www.cst.de/.
15. Kory, C. L., *IEEE Transactions on Electron Devices*, **48**, 38–44 (2001).
16. Visit http://www.apo.saic.com/centers/electro.html.
17. Botton, M., Antonsen, T. M., Jr., Levush, B., Nguyen, K. T., and Vlasov, A. N., *IEEE Transactions on Plasma Science*, **26**, 882–892 (1998).
18. Levush, B., Blank, M., Calame, J., Danly, B., Nguyen, K., Pershing, D., Cooke, S., Latham, P., Petillo, J., and Antonsen, T., Jr., *Physics of Plasmas*, **6**, 2233–2240 (1999).
19. Vlasov, A. N., and T. M. Antonsen, J., *IEEE Transactions on Electron Devices*, **48**, 45–55 (2001).
20. Abe, D. K., Ngo, M. T., Levush, B., Antonsen, T. M., Jr., and Chernin, D. P., *IEEE Transactions on Plasma Science*, **28**, 576–587 (2000).
21. Freund, H. P., and Zaidman, E. G., *Physics of Plasmas*, **7**, 5182–5194 (2000).
22. Herrmannsfeldt, W. B., Electron trajectory program, Tech. Rep. SLAC-R-226, Stanford Linear Accelerator Center (1979), available through http://www.slac.stanford.edu/pubs/.
23. True, R., "A General Purpose Relativistic Beam Dynamics Code," in *1993 Computational Accelerator Physics Conference*, American Institute of Physics, New York, 1994, AIP Conference Proceedings 297, pp. 493–499.
24. Visit http://www.fieldp.com/.
25. Visit http://www.vectorfields.com.
26. Coco, S., Emma, F., Laudani, A., Pulvirenti, S., and Sergi, M., *IEEE Transactions on Electron Devices*, **48**, 24–31 (2001).
27. Visit http://laacg1.lanl.gov/.
28. Visit http://www.ansoft.com/.
29. Cooke, S. J., Mondelli, A. A., Levush, B., Antonsen, T. M., Jr., Chernin, D. P., McClure, T. H., Whaley, D. R., and Basten, M., *IEEE Transactions on Plasma Science*, **28**, 841–866 (2000).
30. Visit http://www.staarinc.com/.
31. Visit htpp://www.ansys.com/.
32. Devanz, G., *Physical Review Special Topics-Accelerators and Beams*, **4** (2001).

5th Workshop on High Energy Density and High Power RF

Snowbird, Utah
October 1-5, 2001

AGENDA

Tuesday – Oct. 2nd

Breakfast
Opening Remarks and Announcements — Michael Fazio
A Sheet Beam Klystron Design *(invited)* — George Caryotakis
High-Power, Annular-Beam Klystron Amplifiers — John Pasour
Klystron Life Results in Particle Accelerator Applications — Heinz Bohlen

Break
100 MW X-Band PPM Focused Klystron — Patrick Ferguson
Multiple Beam Vacuum Electron Device Technology *(invited)* — Edward Wright
Recent Progress in Multi-Beam Klystrons in IECAS — Ding Yaogen
S-Band Multiple-Beam Amplifier Development at NRL — Baruch Levush

Lunch
Inductive Output Tubes-Status and Future Direction *(invited)* — Heinz Bohlen
First Performance of X-Band Pulsed Magnicon Amplifier — Steven Gold
Recent Progress in Understanding the Physics of Plasma-Filled, High-Power Microwave Sources *(invited)* — Gregory Nusinovich
LIGA Microfabrication Overview *(invited)* — Jill Hruby

Break
Discussion: What would revolutionize the power class of linear beam devices? To what extent can novel materials and structures allow operation beyond the TM_{01} mode in devices that classically operate in that mode? — Jeff Calame

Wednesday – Oct. 3rd

Breakfast

"Windowtron" RF Breakdown Studies at SLAC and Snowmass Summary *(invited)*	Lisa Laurent
RF Breakdown in High-Vacuum Multi-Megawatt X-Band Structures	Valery Dolgashev
Active and Passive RF Components for High-Power Systems *(invited)*	Sami Tantawi
Active RF Pulse Compression Using Plasma Switches	Slava Yakovlev

Break

High Power S-Band and X-Band Windows and Water Loads	Patrick Ferguson
Power Supply, Energy Storage Line, and Grid Pulser for High Voltage Gridded Klystron	Roland Koontz
Discussion: Compare microfabricated fundamental mode devices with conventionally fabricated overmoded fast wave devices.	Bruce Carlsten

Lunch
Excursion
Dinner

Thursday – Oct. 4th

Breakfast

Thermal Imaging and Measuring Work Functions of Large Thermionic Cathodes *(invited)*	Cliff Fortgang
Development of Multiple Beam Guns for High Power RF Applications	Lawrence Ives
High-Power Guns for Magnicons	Slava Yakovlev
Development of an X-Band RF Gun at SLAC	Arnold Vlieks

Break

Design and Fabrication of the Klystrino *(invited)*	Glenn Scheitrum
Design and Simulations of Megawatt-Class, MM-Wave Traveling Wave Tubes	Bruce Carlsten
Discussion: Can phased arrays of vacuum electronic devices be made in a tile architecture using microfabrication?	Glenn Scheitrum

Lunch

Fast Wave Devices for Accelerator Applications *(invited)*	Wes Lawson
Theory and Experiment of Ultra High Gain Gyrotron Traveling Wave Amplifier *(invited)*	Kwo Ray Chu
Interaction Circuits for High Average Power Gyro-TWTs Based on Monolithic Lossy Ceramics	Jeff Calame
140 kW, 94 GHz Heavily Loaded TE_{01} Gyro-TWT	David McDermott
Development of a 10 MW, 91 GHz Gyroklystron	Lawrence Ives
Break	
Optimization of Warm Beam CARM-Klystron Efficiency	Steven Gold
Review of Computational Models for HPM Sources *(invited)*	Eric Nelson
The MICHELLE Electron Gun and Collector Modeling Tool	John Petillo
Development of a 3D Finite Element Charged Particle Code with Adaptive Meshing	Lawrence Ives

Friday – Oct. 5th

Breakfast

New Opportunities in Vacuum Electronics Using Photonic Band Gap Structures	Jagadishwar Sirigiri
Progress Toward a Gigawatt-Class Annular Beam Klystron with a Thermionic Electron Gun	Michael Fazio
Relativistic Magnetron with Diffraction Antenna	Mikail Fuks
TBD	TBD
Break	
Discussion: Single beams vs multi-beams vs sheet beams for higher power.	Eric Nelson
Wrap-up and adjourn	Luhmann/Fazio

Author Index

A

Abu-Elfadl, T. M., 45
Antonsen, Jr., T. M., 45
Arfin, B., 159

B

Barnett, L. R., 129
Bliokh, Y. P., 45
Bohlen, H., 21, 35

C

Calame, J. P., 141
Carlsten, B. E., 117, 159
Carmel, Y., 45
Caryotakis, G., 1, 63, 107
Chang, T. H., 129, 147
Chen, C., 151
Chen, S. H., 129
Chu, K. R., 129, 147

D

Danly, B. G., 141
Dolgashev, V. A., 77

E

Earley, L. M., 117

F

Farnham, J., 159
Fazio, M., 159
Fowkes, W. R., 107
Friedman, M., 15
Fuks, M. I., 169

G

Garven, M., 141
Glendinning, F., 63
Goebel, D. M., 45
Granatstein, V. L., 45

H

Haase, A., 159
Habiger, K., 159
Haynes, W. B., 117, 159
Hirata, Y., 147
Hruby, J., 55
Hsu, H. L., 147

J

Jongewaard, E. N., 107

K

Koontz, R. F., 101

L

Landahl, E. C., 107
Laurent, L., 63
Levush, B., 141
Lin, A. T., 147
Loewen, R., 107
Ludeking, L., 15
Luhmann, Jr., N. C., 63, 107, 147

M

McDermott, D. B., 147
Myers, J., 159

N

Nantista, C. D., 83
Nelson, E. M., 159, 177
Nguyen, K. T., 141
Nusinovich, G. S., 45

P

Pasour, J., 15
Pearson, C., 63

S

Schamiloglu, E., 169
Scheitrum, G., 63, 159
Shapiro, M. A., 151
Shkvarunets, A. G., 45
Sirigiri, J. R., 151
Smirnova, E. I., 151
Smith, J., 159
Smithe, D., 15

Song, H. H., 147
Sprehn, D., 63

T

Tantawi, S. G., 77, 83
Temkin, R. J., 151

V

Vlieks, A. E., 107

W

Wheat, R. M., 117

Y

Yaogen, D., 29